T0257790

Crop Production:
A Study of Agricultural Science

Crop Production: A Study of Agricultural Science

Edited by **Corey Aiken**

New York

Published by Callisto Reference,
106 Park Avenue, Suite 200,
New York, NY 10016, USA
www.callistoreference.com

Crop Production: A Study of Agricultural Science
Edited by Corey Aiken

International Standard Book Number: 978-1-63239-133-9 (Hardback)

Printed in the United States of America.

Crop Production: A Study of Agricultural Science

Edited by **Corey Aiken**

New York

Published by Callisto Reference,
106 Park Avenue, Suite 200,
New York, NY 10016, USA
www.callistoreference.com

Crop Production: A Study of Agricultural Science
Edited by Corey Aiken

International Standard Book Number: 978-1-63239-133-9 (Hardback)

Printed in the United States of America.

Contents

Preface

Crop production has always been of quintessence for human civilization. Human beings have always been largely dependent on crops to secure food and fiber. Lately, even fuel has become an essential product derived from crops. There has been a steep rise in the need for crop production in recent times, driven by growing population and changing consumption habits. The purpose of this book is to present the upcoming challenges, advances and future prospects of crop production. It covers extensive subjects like latest agronomic practices to increase productivity, rise of difficulties posed by climatic changes, improvement in use of water resources, usage of fertilizers, etc. This text is ideal for students and experts interested in sustainable crop yield.

This book has been the outcome of endless efforts put in by authors and researchers on various issues and topics within the field. The book is a comprehensive collection of significant researches that are addressed in a variety of chapters. It will surely enhance the knowledge of the field among readers across the globe.

It is indeed an immense pleasure to thank our researchers and authors for their efforts to submit their piece of writing before the deadlines. Finally in the end, I would like to thank my family and colleagues who have been a great source of inspiration and support.

Editor

Irrigation of Sandy Soils, Basics and Scheduling

Mohamed S. Alhammadi and Ali M. Al-Shrouf

Additional information is available at the end of the chapter

1. Introduction

Irrigation of sandy soils must be considered carefully. In this chapter a review is made of the physical characteristics and water-soil relationships of sandy soils [1], as well as various irrigation methods. Recommendations are also given on proper water management at field level.

Dry land and irrigated agriculture depend on the management of two basic natural resources; soil and water. Soil is the supporting structure of plant life and water is essential to sustain plant life. The wise use of these resources requires a basic understanding of soil and water as well as the crop itself.

Irrigating sandy soils requires high attention to the timing and amount of irrigation water applied [2], which are crucial decisions for each operator. Applying too much water means increased pumping costs, reduced water efficiency, and increased potential for pollutant leaching below the rooting zone and into the ground water. Delaying irrigation until plant stress is evident can result in economic yield loss.

This chapter describes some "best" soil moisture management strategies and monitoring techniques that a farmer should consider in managing irrigation and maintaining soil moisture for optimum crop production and least possible degradation of ground water quality.

The main objective of this chapter is to investigate main soil physical and chemical properties that affect the irrigation water amount and to suggest possible irrigation strategies for sandy soil.

Drip irrigation is one of the most efficient irrigation methods to be developed [3]. Although it had been known for many years, some horticulturists and nursery gardeners mostly used it. Its application at the farm level became more common with the extensive use of polyethylene plastics, which has led to reduced cost and increased acceptability for some crops. Trickle irrigation really started in agriculture less than 10 years ago. The real success of this new method is based on a certain number of advantages which are claimed by the enthusiastic promoters of the method: water saving, higher yields, utilization of brackish waters, manual labor extremely reduced, reduction in diseases, weed control, etc. In fact highly qualified specialists have obtained most of these promising results in experimental conditions. Comparative field trials are still few to determine in what proportion these advantages are applicable to large scale irrigation. The method is still in its initial stages and many developments are expected in the near future.

A well designed irrigation scheme may not yield the expected returns if water is not managed in the proper way by farmers. This may be even more applicable in the case of sandy soils for which irrigation must be handled with special care. The human aspect is often unduly disregarded during the planning period whereas it plays a decisive part during the whole lifetime of a project. It is therefore necessary to provide farmers with the knowledge that they need. This of course can only be done through an intensive education program of demonstrations, advice, rewards, etc., carried out by a well organized extension service. In order to be efficient, the extension workers should receive special training on the irrigation of sandy soils based on a very good knowledge of the local soil conditions.

2. Basic soil characteristics affecting irrigation

Soil is composed of three major parts: air, water, and solids. The solid component forms the framework of the soil and consists of mineral and organic matter [4]. The mineral fraction is made up of sand, silt, and clay particles. The proportion of the soil occupied by water and air is referred to as the pore volume. The pore volume is generally constant for a given soil layer but may be altered by tillage and compaction. The ratio of air to water stored in the pores changes as water is added to or lost from the soil.

There are two main soil characteristics that affect irrigation:

- Soil physical characteristics [5] (soil depth, soil texture, soil infiltration, soil moisture content, bulk density and soil porosity). Summary of the physical characteristics of the main soil texture classes are listed in table 1.

- Soil chemical characteristics [6] (soil salinity and sodicity)

2.1. Soil depth

Soil depth refers to the thickness of the soil materials, which provide structural support, nutrients, and water for plants. Sandy lands are deep and contain low gravels at a depth of more than 50 centimeters. These lands are characterized by high quantities of calcium carbonate and found in some area of the gravel plain adjacent to the sandy desert. The depth of the soil layer of sand and gravel can affect irrigation management decisions. If the depth to this layer is less than 3 feet, the rooting depth and available soil water for plants is decreased. Soils with less available water for plants require more frequent irrigation.

2.2. Texture

The distribution of the soil particles according to their size is called the texture. Sand is usually defined as particles having a diameter between 0.05 and 1.00 mm (Figure 1). If the amount of particles within this range is greater than 50 percent, the soil is said to be sandy. According to the exact percentages of sand and other particles contained in a sandy soil, its texture will vary from sandy-clay to coarse sand (Figure 2). The texture of sandy soils has a very important influence on the infiltration rate, the water holding capacity and consequently on its value for irrigation. Since generally the more loam and clay contained in the soil, the better it will be for irrigation, a mechanical analysis is a necessary tool for soil classification. Coarse soils are easily eroded by running water, which is one of the obstacles to successful surface irrigation. Two well known physical features of sandy soils are their coarse texture and their high rate of permeability. Other features that play an important part in irrigation are the pore space, the bulk density and the water content [7].

Since sand particles are most dominant in the soil of arid regions, their texture tends to be light (sandier). Moreover, clay particles are rarely found in some areas of the arid regions with coarse sand and large quantity of gravel. Thus fast leakage of nutritional elements occurs when water is added at short intervals. Adding organic substances is highly recommended to improve the soil water holding capacity.

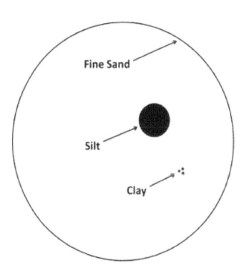

Figure 1. Comparative size of clay to fine sand. Clay is actually less than 0.002 mm with fine sand up to 0.25 mm [8].

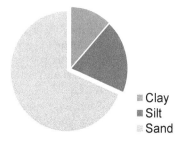

■ Clay
■ Silt
■ Sand

Figure 2. Percentages of sand and other particles contained in a sandy soil vary from sandy clay to coarse sand.

2.3. Soil infiltration

The infiltration rate is the velocity at which water percolates into a soil and usually decreases the longer water is in contact with the soil. It will reach a relatively steady value equal to the permeability or hydraulic conductivity of water through the soil. This variation of infiltration rate with time differs from one type of soil to another. In the case of sandy soils the final rate is reached rapidly and is usually high.

Sandy soils have high infiltration rates varying for sandy clay and sandy loam from 4 to 25 cm/h, but in very permeable sandy soils values as high as 100 to 400 cm/h are easily reached (Figure 3 & 4). High final infiltration rates are responsible for important water losses both in conveyance systems and in the field. Soils having a final infiltration rate of 10 cm/h and above are generally not recommended for surface irrigation systems. In other words, to keep the conveyance and application efficiencies at an acceptable level the length of the ditches and the size of the fields may be too small for proper cropping. A 30 l/s flow could irrigate no more than 1080 m² at any one time. The uniformity of application will be poor if the fields are large since upper parts would receive more water than the lower ones. High infiltration rates may be an important constraint to efficient surface irrigation schemes. High infiltration rates also have an action on the soil structure in that often the clay particles contained in the upper layers are conveyed to deeper layers where they accumulate and form a less permeable horizon. This horizon may impede the deep percolation of excess water coming from rain or over-irrigation and may form a perched water table, which will require field drainage.

Infiltration is the downward entry of water into the soil. The velocity at which water enters the soil is the infiltration rate. The infiltration rate is typically expressed in inches per hour. Water from rainfall or irrigation must first enter the soil to be valuable for the plant. This is important in irrigation since infiltration is considered as an indicator of the soil's ability to allow water movement into and through the soil profile. Soil temporarily stores water, making it available for root uptake, plant growth and habitat for soil organisms. When water is supplied at a rate that exceeds the soil's infiltration capacity, it moves down slope as runoff on sloping land or ponds on the surface of level land. When runoff occurs on bare or poorly vegetated soil, erosion takes place. Runoff carries nutrients, chemicals, and soil with it, resulting in decreased soil productivity, off-site sedimentation of water bodies and diminished

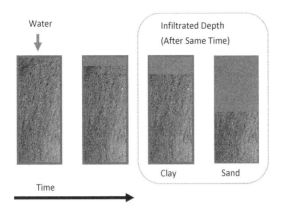

Figure 3. Comparison of water infiltrated depth over time between two soil textures.

water quality. Sedimentation decreases storage capacity of reservoirs and streams and can lead to flooding. Therefore, knowing of soil infiltration rate is important to select a good method of irrigation.

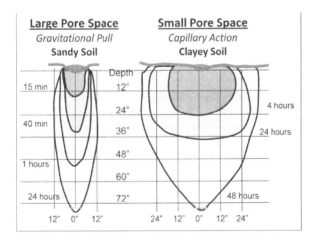

Figure 4. Comparative movement of water in sandy and clayey soils. In sandy soils, water readily moves downward due to the gravitational pull. In clay soils, water slowly moves in all direction by capillary action [8].

2.4. Soil moisture content

Soil moisture content is one of the most variable characteristics of soil. The soil acts as a reservoir for water, making it available for plants as it is needed. Soil water is very important to the entire soil system, not only because it is necessary for plant growth, but because the nutrients required for plant growth are also present in the soil solution. Most of the important soil reactions (weathering, cation exchange, organic matter decomposition, fertilization) take place in the context of the soil solution. Thus, soil moisture is a key property of the soil.

2.5. Bulk density and porosity

The porosity or pore space is that space between the soil particles, which is equal to the ratio of the volume of voids either filled with air or with water to the total volume of soil, including air and water. The porosity or pore space of sandy soils is less than for clay soils.

The apparent specific gravity or bulk density is the ratio of the weight of a given volume of dry soil (dry space included) to the weight of an equal volume of water. The apparent specific gravity varies with soil types as does the porosity. Mean and extreme values are indicated in table 2.

Properties	Sandy	Loamy	Clay
Feeling	Coarse	Medium	Fine
Decomposition of O.M.	Rapid	Medium to high	High
Water Holding Capacity	Low	Medium to high	High
Drainage	Good	Medium	Poor
Aeration	Good	Medium	Poor
Infiltration rate	High	Medium	Low
Porosity	Low	Medium	High

Table 1. Physical characteristics of main soil texture classes [4].

Soil Texture	Apparent Specific Gravity	Pore Space
Sandy		
Mean	1.65	38
Extremes	1.55-1.80	32-42
Sandy loam		
Mean	1.50	43
Extremes	1.40-1.60	40-47

Table 2. Bulk density and porosity of some sandy and sandy loam soil

2.6. Soil salinity and sodicity

Salinity in soil becomes a problem when the total amount of salts that accumulate in the root zone is high enough to negatively affect plant growth. Excess soluble salts in the root zone restrict plant roots from withdrawing water from the surrounding soil, effectively reducing the plant available water. Plant available water is at its maximum and soil salinity is at its lowest concentration immediately after irrigation. However, as plants use soil water, the force with which the remaining water is held in the soil increases, making it progressively more difficult to withdraw water. Also, as water is taken up by plants through transpiration or is lost to the atmosphere by evaporation, the salinity of the remaining water increases. This is due to the fact that the majority of the salts are left behind in both processes while the amount of water that the salt is dissolved in is progressively reduced. This effect becomes most pronounced during periods of high evapotranspiration (ET) demand, such as hot sunny summer days and during the peak of the growing season. It is widely accepted that the salinity of soil water is equal to approximately three times the salinity of irrigation water, assuming relatively little leaching is occurring [5]. In conditions of relatively high leaching fractions, the soil water solution and drainage water will have a salinity level slightly greater than the irrigation water. When considering salinity effects of the irrigation water, the plants and soil are actually subject to the salinity of the resultant soil solution, which is a function of the salinity of the applied water.

In much of the arid and semi-arid regions, most of the salts present in irrigation water and groundwater are either chlorides, sulfates, carbonates or bicarbonates of calcium, magnesium, sodium and potassium. Each of these salts has a unique solubility, which along with the composition of the mineral material through which water passes, dictates the salts present in the water. When these salts are dissolved in solution, they often ionize, breaking down (disassociating) into cations (positively-charged molecules) and anions (negatively-charged molecules) [6]. The most common cations in arid and semi-arid areas are calcium, magnesium and sodium. Each of these cations is base-forming, meaning that they contribute to an increased OH^- concentration in the soil solution and a decrease in H^+ concentration. They typically dominate the exchange complex of soils, having replaced aluminum and hydrogen. Soils with exchange complexes saturated with calcium, magnesium and sodium have a high base saturation and typically high pH values

In addition to decreasing plant available water and being potentially toxic to plants, soil solution salinity can also affect soil physical properties. Salinity can have a flocculating affect on soils, causing fine particles to bind together into aggregates.

Assessment of the relationship between soil solution salinity and soil physical properties requires knowledge of the constituents of the dissolved salts, and especially the sodium concentration [9]. Sodium has the opposite effect on soils that salinity does. While elevated electrolyte concentration may enhance flocculation, sodium saturation may cause dispersion. Because of its relatively large size, single electrical charge and hydration status, adsorbed sodium tends to cause physical separation of soil particles.

The relationship between soil salinity and its flocculating effects, and soil ESP (exchangeable sodium percentage) and its dispersive effects, dictate whether or not a soil will stay aggregated or become dispersed under various salinity and sodicity combinations

Dispersed clay particles within the soil solution can clog soil pores when the particles settle out of solution. Additionally, when dispersed particles settle, they may form a nearly structureless cement-like soil. This pore plugging and cement-like structure make it difficult for plants to get established and grow. It also impedes water flow and water infiltration into the soil.

The disruption of soil hydraulic properties has two main consequences. Firstly, there is less water infiltrating into the soil, and therefore less plant available water, particularly at deeper depths. Secondly runoff, and therefore water loss and soil erosion, may be enhanced, and both affect the irrigation.

3. Levels of soil moisture content

There are four important levels of soil moisture content that reflect the availability of water in the soil (figure 5). These levels are commonly referred to as: 1) saturation, 2) field capacity, 3) wilting point and 4) oven dry [4]

Saturated: all of the soil pore spaces are filled with water. At this point the soil is at its maximum retentive capacity. Not all the water present in the soil is available for plant use. Some water drains past the rooting zone and is unavailable. Soil can be viewed as a sponge composed of air and solid particles when dry. This is an undesirable condition for the growth of most plants because the available dissolved oxygen is quickly depleted. Water at the saturation point in soils is held at a tension of 0 MPa (0 bars).

Field capacity: the maximum amount of water left in the soil after losses of water to the forces of gravity have ceased and before surface evaporation begins. It occurs when the soil contains the maximum amount of capillary water. Field capacity represents the amount of water remaining in the soil after the large pores have drained. Medium and small pores are still filled with water held against the force of gravity. Soil water at field capacity is readily available to plants and sufficient air is present for root and microbial respiration. The optimum water content for plant growth and soil microbial respiration is considered to be close to field capacity. Soils at field capacity are generally considered to be holding water at a tension of about 0.01-0.03 MPa (0.1-0.3 bars).

Permanent wilting point (PWP): If water is continually taken-up by plants and no additional water is added to the soil in the form of precipitation or irrigation water, the medium and small soil pores will be emptied of water. With time, the plant will eventually wilt when it cannot extract more water. The soil is said to be at the permanent wilting point when plants can no longer exert enough force to extract the remaining soil water. At the permanent wilting point, water is held in the soil at about 1.5 MPa (15 bars). At this point the plant can no longer obtain sufficient water from the soil to meet its transpiration needs. At this point the plant enters permanent wilt and dies.

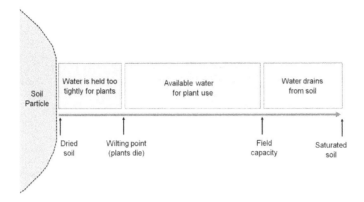

Figure 5. arrangement of soil water classes starting from the surface of the soil particle.

Hygroscopic coefficient: The hygroscopic coefficient is the boundary between moist-appearing and dry-appearing soil. This boundary is not sharp. Its arbitrary value is – 3100 kPa (-3.1 Mpa) soil matric potential.

Oven-dry: If soil is placed in an oven and dried at 105°C, additional water will be removed. The oven-dry condition is the reference state used as the basis for expressing most soil character-istics.

3.1. Classes of soil moisture

Gravitational water: The water that moves into, through, or out of the soil by gravity (0- 1/3 bar).

Capillary water: Water that is left in soil, along with hygroscopic moisture and water vapor, after the gravitational water has drained off. Capillary water is held by surface tension as a film of moisture on the surface of soil particles and peds, and as minute bodies of water filling part of the pore space between particles. Most, but not all, of this water is available for plant growth Capillary water is held in the soil against the pull of gravity forces acting on capillary water (1/3 to -31 bars)

Hygroscopic water: Water absorbed from the atmosphere and held very tightly by the soil particles, so that it is unavailable to plants in amounts sufficient for them to survive. It is the water lost from an air-dry soil when it is heated to 105°C (31-10,000 bar).

The water content of soil depends on the amount of water stored in the pore space. Since sandy soils do not have a very large total pore space, their water content will never be very high. The water content can be expressed either as a percentage on a dry weight basis or as a percentage on a volume basis. The moisture content on a dry weight basis is equal to the ratio of the weight of water contained by a soil at field capacity to the weight of this same soil after having been dried in an oven at a temperature of 105 °C. The moisture content on a volume basis is equal to the ratio of the volume occupied by the water stored in the pore space to the total volume of the soil.

This relation can be converted to a depth of water available in soil which is usable in determining quantitative water requirements of individual crops grown in specific soils and is called the readily available water.

In irrigation practice it is not recommended to wait for the soil water to reach the permanent wilting point before replenishing the soil reservoir. In this way the plants will not suffer from an eventual lack of water. In principle, water applications should never be greater than the readily available water as any excess will automatically be lost by deep percolation.

3.2. Available water for various sand soil textures

The more sandy a soil is the higher its apparent specific gravity and the lower its pore space. Thus sandy soils, which are often designated by farmers as "light soils" are in fact the ones that weigh the most per unit volume. The term "light" refers to their ease in working with agricultural implements.

Available soil water defined as the amount of water present in a soil, which can be moved by plants. It is designated as the difference between the field capacity and the wilting point. The available water for various sandy soil classes is listed in table 3.

Soil Texture	Available Water (mm/m)
Coarse Sands	20-65
Fine Sands	60-85
Loamy Sands	65-110
Sandy Loams	90-130
Fine Sandy Loam	100-170

Table 3. Available water for various sandy soil textures [10]

4. Irrigation of sandy soils

There are about 250 million irrigated hectares in the world, most of which have experienced moderate to high soil erosion due to the lack of experience [10]. Surface runoff and deep percolation decrease the irrigation efficiency; therefore selecting suitable irrigation method is needed in order to improve irrigation efficiency. More than 95% of irrigated land in the world is irrigated by surface method. Using surface irrigation method is not desirable on sandy soil due to the fact that it has high infiltration rate. On the other hand, sprinkle and drip irrigation methods are more desirable on the sandy soils because surface runoff and deep percolation do not occur frequently. In the drip irrigation system, water is applied more to the places of the plant root area. Compare to other irrigation methods, drip irrigation is more efficient and can be applied more frequently. The low irrigation rate of drip irrigation makes it most suitable for both sandy and clay soils. In addition, drip irrigation method has been successfully used with saline water.

5. Irrigation scheduling

As the demand for water increases, along with the need to protect aquatic habitats, water conservation practices for irrigation need to be effective and affordable. Precision irrigation will optimize irrigation by minimizing the waste of water, and energy, while maximizing crop yields [11].

The most effective method for determining the water demands of crops is based on the real time monitoring of soil moisture, and direct water application used in conjunction with the information about soil hydrological properties and evapotranspiration [12].

Good irrigation scheduling means applying the right amount of water at the right time. In other words, making sure water is available when the crop needs it. Scheduling maximizes irrigation efficiency by minimizing runoff and percolation losses. This often results in lower energy and water use and optimum crop yields, but it can result in increased energy and water use in situations where water isnot being managed properly.

The following factors influence an efficient irrigation schedule:

1. *Irrigation System Information:* A key piece of irrigation system information is the net amount of water applied by your system. Determining this requires measurements of system gross application rate and irrigation application efficiency. In addition to the application efficiency, it is very important to consider the uniformity efficiency especially in the case of sprinkler irrigation system in windy areas.

2. *Crop Information:* crop type, crop density, amount of vegetative cover or leaf area, crop health, and stage of growth

3. *Climatic conditions:* temperature, relative humidity, wind speed, day time hour, solar radiation

4. *Environmental Conditions:* Soil salinity [3], soil depth and layering, poor soil fertility

5. *Field Management Practices.* Fertility management, disease and pest control

Advantages of irrigation scheduling

Irrigation scheduling offers several advantages:

1. It enables the farmer to schedule water cycles among the various fields to minimize crop water stress and maximize yields.

2. It lowers fertilizer costs by holding surface runoff and deep percolation (leaching) to a minimum.

3. It increases net returns by increasing crop yields and crop quality.

4.   It assists in controlling root zone salinity problems through controlled leaching.

5.   It results in additional returns by using the "saved" water to irrigate non-cash crops that otherwise would not be irrigated during water-short periods.

5.1. Irrigation scheduling metchods

These methods can range from *qualitative 'feeling'* the need to apply water to prevent damaging crop stress to *quantitative soil water budgets* for individual fields at various levels of detail. Table (4) describes the main technologies to improve irrigation efficiency.

Irrigation Scheduling Methods	Method Characteristics	Level of Complexity
Visual observations	• Crop is at a specific growth stage • Crop is experiencing water stress • Water is available to irrigate • Neighbor is irrigating	Simple
Measure current moisture level	• Soil moisture sensors indicate water availability in the root zone e.g. gypsum blocks, granular matrix probes, tensiometers, capacitance probes, • Sensors indicating the water status of the crop e.g. IR thermometer for canopy temperature [13] and [14]	
Smart Irrigation: Defined as controllers that reduce water use by monitoring and using information about site conditions (such as soil moisture, rain, wind, slope, soil, plant type, and more and applying the right amount of water based on those factors". Essentially, these irrigation controllers receive feedback from the irrigated system and schedule or adjust irrigation duration and/or frequency accordingly [15].	Maintain soil water budget of the root zone (Weather-Based Irrigation Control System Principles) • Daily measurement of the actual ET using lysometers • Daily reference ET based on weather data • Crop coefficients for actual ET • Measure rainfall • Forecast daily water use for next 1-2 weeks • Compare computed with measured soil moisture Automatic irrigations (Smart Irrigation System) • measured soil moisture level in root zone • water potential of the plant • computed soil water depletions • programmed time	Basic to Moderate Methods
Infra red thermometer: Canopy temperature is measured with infrared thermometer. It also simultaneously measures canopy temperature and air temperature and displays the difference. This difference can be used for scheduling irrigation. [16]	• If canopy temperature "/> Air Temperature: indicates stress irrigation is scheduled • Canopy Temperature < Air Temperature: indicate the plants have sufficient amount of water	
Remote sensing: In areas, where a single crop is grown [17]	Irrigation scheduling can be done with the help of remote sensing data. Reflectance of solar radiation by the plants with sufficient amount of water is different from that of stressed plants. This principle can be used for scheduling irrigation.	(Advance Method) Complex

Table 4. main technologies to improve irrigation efficiency

The irrigation scheduling method can be classified over 5 levels as follows:

Level	Irrigation Scheduling Method
0	"Feel Like It" Method (Guessing)
1	"Feel and See" Method
2	Use a Schedule (1/2" every 3 days)
3	Use a Soil Water Tension device
4	Use a Soil Water Tension device to apply water on a schedule
5	Adjust water based on crop need, utilizing Soil Water Tension device

Table 5. classification of irrigation scheduling methods [18]

6. Irrigation water management

Irrigation Water Management is the process of determining and controlling the volume, frequency, and application rate of irrigation water in a planned [19], efficient manner. Irrigation water management is important for the following reasons

- Managing soil moisture to promote desired crop response.
- Optimizing the use of available water supplies.
- Preventing economic yield losses due to moisture stress.
- Maximizing efficiency of production inputs.
- Minimizing irrigation induced erosion.
- Minimizing leaching potential of nitrates and other agrichemicals below the rooting zone.
- Managing salts in the crop root zone.
- Managing air, soil or plant micro-climate.

Irrigation water management primarily aims to control the volume and frequency of irrigation water applied to crops, so as to meet crop needs while conserving water resources. This includes determining and controlling the rate, amount and timing of irrigation water in a planned and efficient manner. Obtaining high yield from irrigation requires appropriate management of all the inputs. This means efficient management of the amount of applied water. One of the most important aspects of irrigation management is deciding when to turn on the irrigation system and how much water to apply [20]. Fortunately, irrigation scheduling methods have been developed to help making those decisions. Using rational or scientific methods to schedule irrigations is essential for good irrigation management. Good irrigation management begins with accurate measurement of the rainfall received and knowing the soil moisture status in the field. Over the past few years, several attempts were conducted to develop irrigation scheduling methods [21]. Measurement of soil moisture levels is the most common method in determining irrigation scheduling, however more modern methods were developed that combine both crop water use and soil water estimate. The major factors influencing irrigation management in addition to the management levels will be discussed in this section.

Five essential elements for irrigation water management:

$$\left.\begin{array}{l} 1.\text{Measurement of soil moisture} \\ 2.\text{Measurement of irrigation water applied} \\ 3.\text{Measurement crop consumptive use} \\ 4.\text{Evaluation of the irrigation system efficiency} \\ 5.\text{Measurement of the soil and irrigation water quality} \end{array}\right\} \text{Need all 5 items} \qquad (1)$$

Table 6 shows a comparison between the currently used smart irrigation technology and the traditional controller, which have been developed to address some elements of irrigation water management to improve irrigation efficiencies.

Features	Traditional controller	"Smart" controller
Automated watering system	X	X
Does not require seasonal monitoring/changes		X
Uses less water		X
Fails to consider; Rainfall, Solar Radiation, Humidity, Temperature and Wind	X	

Table 6. comparison between traditional controller and smart controller [18]

7. Irrigation using wireless sensors

Nowadays new technologies are being used to determine the actual crop water need through sensing soil moisture using wireless techniques (figure 6). These technologies have been used in precision agriculture to assist in precise irrigation where it can provide a potential solution to efficient water management through remotely sensing soil water conditions in the field and controlling irrigation systems on the site [22]. The network of these sensors consists of radio frequency transceivers, sensors, and microcontrollers. There are several types of wireless technologies, most of which depend on infrared light, point-to-point communications, wireless personal area network (WLAN), Bluetooth, ZigBee, multi-hop wireless local area network, and log-distance cellular phone systems such as GSM/GPRS and CDMA [23]. The technology of wireless sensors needs more development especially in the agricultural area. Since the late 1970s, a wide range of competing technologies has each been hailed as 'the answer' for sensing soil water. Several studies were conducted to evaluate these technologies [24], [25], [26] and [27]. A new wireless irrigation system was developed that communicates to boost irrigation through Bluetooth and wireless network of small soil moisture and temperature sensors. It has the ability to continuously detect accurate plant water needs [24]. Chávez et al. [25] used a new system with Single Board Computer (SBC) using Linux operating system to control solenoids connected to an individual or group of nozzles. The control box was connected to a sensor network radio, GPS unit, and Ethernet radio. For efficient irrigation water management, Kim and Evans [27] designed a Bluetooth wireless communication in-field sensor and control software using four major design factors that provide real-time monitoring and control of both field data and sprinkler controls. The system successfully enabled real-time remote access to the field conditions and site-specific irrigation. Their recommendation was to simply use and manage the software by growers. Smart sensor array for cotton irrigation scheduling was developed by Vellidis et al. [28]. Damas et al [24] developed and evaluated a remotely controlled automatic irrigation system for an area of 1500 ha using WLAN network. They were able to save 30-60% of water usage. In addition, they develop a mobile field data acquisition system to collect information for crop management and spatial-variability studies including available soil water and plant water status and other field data. Evans and Bergman [29] assisted irrigation scheduling using wireless sensors in three irrigation systems (e.g. self-propelled, linear-move and center-pivot) combined with an on-site weather station, remote sensing data and farmer preferences.

Since wireless sensors and networks have just started in the agricultural field, there is great potential for the implementation of wireless sensors in the irrigation of sandy soil, in which greater water use efficiency will be achieved. The main advantage of these technologies is to save water use in agriculture sector.

Figure 6. wireless sensor in the field that detects soil water status at various depths.

8. Conclusion

Sandy soils have a low pore space and a high infiltration rate. The consequences of these two features on irrigation systems and methods are of paramount importance. The low pore space is responsible for a low water holding capacity. Consequently the frequency of irrigation and the labor requirements are high regardless of the irrigation method used. Labor requirements can be reduced but the initial cost of equipment is then considerably increased. The high infiltration rates make surface irrigation very difficult, as an important task is to avoid losses when applying water to the fields. The adaptation of surface irrigation, when possible, requires higher investment costs for increased length and size of canals, canal lining, large number of small plots and eventually special on-farm equipment.

On the contrary high infiltration rates have little influence on sprinkler irrigation. This method can therefore be considered as the best for sandy soils. It will lead to acceptable efficiencies if properly designed and managed.

Drip irrigation is a promising method but its cost is still quite high. It is recommended to set up field trials before embarking on large scale developments with drip irrigation.

Wireless sensors and networks have just started in agriculture to assist in precise irrigation where it can provide a potential solution to efficient water management through remotely sensing soil water conditions in the field and controlling irrigation systems in the site. The implementation of wireless sensors in the irrigation of sandy soil has a great potential, in which more water use efficiency will be achieved. More development of such technology is needed specifically for the agricultural area.

Author details

Mohamed S. Alhammadi and Ali M. Al-Shrouf

Research & Development Division, Abu Dhabi Food Control Authority, United Arab Emirates

References

[1] Neal J.S., S.R. Murphy, S. Harden, W.J. Fulkerso. Differences in soil water content between perennial and annual forages and crops grown under deficit irrigation and used by the dairy industry. Field Crops Research, Volume 137, 2012, Pages 148-162.

[2] Sánchez N., J. Martínez-Fernández, J. González-Piqueras, M.P. González-Dugo, G. Baroncini-Turrichia, E. Torres, A. Calera, C. Pérez-Gutiérrez. Water balance at plot scale for soil moisture estimation using vegetation parameters. Agricultural and Forest Meteorology, Volumes 166–167, 2012, Pages 1-9.

[3] Mei-xian LIU, Jing-song YANG, Xiao-ming LI, Mei YU, Jin WANG. Effects of Irrigation Water Quality and Drip Tape Arrangement on Soil Salinity, Soil Moisture Distribution, and Cotton Yield (Gossypium hirsutum L.) Under Mulched Drip Irrigation in Xinjiang, China. Journal of Integrative Agriculture, Volume 11, Issue 3, March 2012, Pages 502-511.

[4] Brady, N.C., and R.R. Weil. The nature and properties of soils (14th ed.). 2007. Prentice Hall, Upper Saddle River, NJ.

[5] Aragüés R., V. Urdanoz, M. Çetin, C. Kirda, H. Daghari, W. Ltifi, M. Lahlou, A. Douaik. Soil salinity related to physical soil characteristics and irrigation management in four Mediterranean irrigation districts. Agricultural Water Management, Volume 98, Issue 6, April 2011, Pages 959-966.

[6] Tedeschi A., R. Dell'Aquila Effects of irrigation with saline waters, at different concentrations, on soil physical and chemical characteristics. Agricultural Water Management, Volume 77, Issues 1–3, 22 August 2005, Pages 308-322.

[7] Seyed Hamid Ahmadi, Finn Plauborg, Mathias N. Andersen, Ali Reza Sepaskhah, Christian R. Jensen, Søren Hansen. Effects of irrigation strategies and soils on field grown potatoes: Root distribution. Agricultural Water Management, Volume 98, Issue 8, 30 May 2011, Pages 1280-1290.

[8] Whiting, D., Card, A., Wilson, C. Moravec, C., Reeder, J.. Managing Soil Tilth, Texture, Structure and Pore Space. Colorado Master Gardner Program 2011, Colorado State University Extension. CMG GardenNotes #213.

[9] Al-Shrouf A. "The safe use of marginal quality water in agriculture, challenges and future alternative, I-Saline water" a paper presented at The Sixth Annual UAE University Research Conference at the United Arab Emirates University. April 24th-26th, 2005, Al-Ain.UAE.

[10] Hargreaves, G.H. and Merkley, G.P.. Irrigation Fundamentals. Water Ressources Publications 1998. Colorado. USA

[11] Seyed Hamid Ahmadi, Mathias N. Andersen, Finn Plauborg, Rolf T. Poulsen, Christian R. Jensen, Ali Reza Sepaskhah, Søren Hansen. Effects of irrigation strategies and soils on field grown potatoes: Yield and water productivity. Agricultural Water Management, Volume 97, Issue 11, 1 November 2010, Pages 1923-1930

[12] Algozin K. A., V. F. Bralts and J. T. Ritchie. Irrigation scheduling for a sandy soil using mobile frequency domain reflectometry with a checkbook method. Journal of Soil and Water Conservation 2001 56(2):97-100

[13] Cardenas B. -Lailhacar, M.D. Dukes Precision of soil moisture sensor irrigation controllers under field conditions. Agricultural Water Management, Volume 97, Issue 5, May 2010, Pages 666-672.

[14] Nolz R., G. Kammerer, P. Cepuder. Calibrating soil water potential sensors integrat-
      ed into a wireless monitoring network. Agricultural Water Management, Volume
      116, 1 January 2013, Pages 12-20.

[15] McCready M.S., M.D. Dukes, G.L. Miller. Water conservation potential of smart irri-
      gation controllers on St. Augustinegrass. Agricultural Water Management, Volume
      96, Issue 11, November 2009, Pages 1623-1632.

[16] Padhi J., R.K. Misra, J.O. Payero. Estimation of soil water deficit in an irrigated cotton
      field with infrared thermography. Field Crops Research, Volume 126, 14 February
      2012, Pages 45-55.

[17] Folhes M.T., C.D. Rennó, J.V. Soares. Remote sensing for irrigation water manage-
      ment in the semi-arid Northeast of Brazil. Agricultural Water Management, Volume
      96, Issue 10, October 2009, Pages 1398-1408.

[18] Al-Shrouf A. 2008 "Irrigation Requirements of Date Palm (Phoenix dactylifera) in
      UAE Conditions" a paper presented at the ninth Annual UAE University Research
      Conference at the United Arab Emirates University. April 21th-23th, Al-Ain.UAE.

[19] Thompson R.B., M. Gallardo, L.C. Valdez, M.D. Fernández. Using plant water status
      to define threshold values for irrigation management of vegetable crops using soil
      moisture sensors. Agricultural Water Management, Volume 88, Issues 1–3, 16 March
      2007, Pages 147-158.

[20] Jorge A. Delgado, Peter M. Groffman, Mark A. Nearing, Tom Goddard, Don Reicos-
      ky, Rattan Lal, Newell R. Kitchen, Charles W. Rice, Dan Towery, and Paul Salon.
      New technology to increase irrigation efficiency. Journal of Soil and Water Conserva-
      tion 2008 63(1):11A; doi:10.2489/jswc.63.1.11A

[21] Shock, C.C., A.B. Pereira, B.R. Hanson, and M.D. Cahn. 2007. Vegetable irrigation. p.
      535--606. In R. Lescano and R. Sojka (ed.) Irrigation of agricultural crops. 2nd ed.
      Agron. Monogr. 30. ASA, CSSA, and SSSA, Madison, WI.

[22] Richard E. Plant. Expert systems in agriculture and resource management. Techno-
      logical Forecasting and Social Change, Volume 43, Issues 3–4, May–June 1993, Pages
      241-25.

[23] Wang, N., Zhang, N., Wang, M.. Wireless Sensors in Agriculture and Food Industry -
      Recent Development and Future Perspective. Computers and lectronics in Agricul-
      ture 2006. Volume 50, 1:1-14.

[24] Kim, Y., Evans, R. G., Iversen, W.M.. Remote Sensing and Control of an Irrigation
      System Using a Distributed Wireless Sensor Network. IEEE Transaction on Instru-
      mentation and Measurement 2008. Volume 57, 7:1379-1387.

[25] Chávez, J.L., Pierce, F.J., Elliott, T.V. and Evans, R.G. A Remote Irrigation Monitoring
      and Control System for Continuous Move Systems. Part A: Description and Develop-
      ment. Precision Agriculture 2009. Volume 11, 1:1-10.

[26] Damas, M., Prados, A.M., Gómez, F., Olivares, G.. HidroBus® System: Fieldbus for Integrated Management of Extensive Areas of Irrigation Land. Microprocessors Microsyst 2001. 25:177-184.

[27] Kim, Y. and Evans, R.G.. Software design for wireless sensor-based site-specific irrigation. Computers and Electronic in Agriculture 2009. Volume 66, 2:159-165.

[28] Vellidis, G., Tucker, M.; Perry, C.; Wen, C.; Bednarz, C.. A real-time wireless smart sensor array for scheduling irrigation. Comput. Electron. Agric. 2008. 61, 44-50.

[29] Evans, R. and Bergman, J. 2003. Relationships Between Cropping Sequences and Irrigation Frequency under Self-Propelled Irrigation Systems in the Northen Great Plains (Ngp). USDA Annual Report. Project NO. 5436-13210-003-02. June 11, 2003 – Dec. 31, 2007.

Chapter 2

Development of an Environmentally Advanced Basin Model in Asia

Kazuo Oki, Keigo Noda, Koshi Yoshida,
Issaku Azechi, Masayasu Maki, Koki Homma,
Chiharu Hongo and Hiroaki Shirakawa

Additional information is available at the end of the chapter

1. Introduction

Tropical regions support a large number of plant and animal species, and conservation of these regions is a major issue that must be tackled globally, not only by the nations in tropical regions. Agriculture has a major impact on the environment in tropical regions, including Asian nations, which face four issues in relation to the environment.

The first issue is the expansion of cultivated land and the accompanying increase in water demand. At issue is land use and production planning to accommodate the increase in food demand accompanying population growth.

The second issue is environmental problems resulting from the spread of modern agricultural methods. Since the Green Revolution, Asian nations have greatly increased land productivity through the widespread use of modern agricultural methods such as the adoption of high-yielding varieties and chemical fertilizers, in response to population growth. However, modern agricultural methods that promote uniform cultivation simplify ecosystems and are harmful to regional biodiversity. This in turn erodes regional characteristics and weakens the ability of regions to adapt to external change, which carries the risk that a major environmental change could result in catastrophic damages.

The third issue is the increase in the demand for biomass energy. While biomass energy is expected to be used more extensively to reduce the use of fossil fuels and lower carbon dioxide (CO_2) emissions, there are fears that the expansion of cultivated land for the production of biomass energy crops will reduce forest areas.

The fourth issue is the concern that global warming will lead to a decrease in agricultural productivity. The Fourth Assessment Report from the Intergovernmental Panel on Climate Change (IPCC) examines the impact of global warming on food production and predicts that in low-altitude regions, particularly in tropical regions with dry and rainy seasons, a rise of just one to two degrees Celsius in regional temperatures will lower crop productivity and increase the risk of famine.

To mitigate these issues, it is desirable to develop and disseminate an environmentally advanced model in Asia that takes into consideration the balance of water, food and energy in response to climate change. At the same time, native varieties that are effective, together with native cultivation methods and traditional methods of using local resources that are effective in developing Asian nations, should be actively used. Furthermore, it is necessary to conduct reliable research to find ways to achieve economic betterment through agriculture, and consider the planning and dissemination of technological developments that incorporates the three factors of water, food production and energy at the basin level.

We chose to study the Citarum River Basin, located in the Cianjur Regency in West Java of the Republic of Indonesia. Cianjur is located midway between the cities of Bandung and Bogor, and is situated upstream of the Citarum River Basin. The basin includes the Jatiluhur Dam and Reservoir, which is a reservoir for the capital of Jakarta. The region is a belt for the production of highly palatable rice; however, sediment accumulation and eutrophication have become serious issues in downstream reservoirs, due to the inflow of waste water from urban areas and fertilizer components from hilly upland fields into rivers. In 2009, the MNN (Mother Nature Network) portal for environmental news identified the region as being the most polluted in the world.

To develop and propose an environmentally advanced basin model in Asia, we performed the following tasks: (1) Assess flood risks, drought risks and nitrogen loads; (2) assess the food production potential with remotely sensed data; (3) identify the lands that are suitable for the development of a wide-area assessment model to predict rice growth and yields that takes into consideration weather conditions and variety characteristics; (4) estimate the supply and demand for biomass energy in Indonesia; and (5) propose poverty alleviation policy for farmers as an example of an environmentally advanced basin model in Asia with a focus on water, food and energy.

Figure 1 show the study groups for proposing the environmentally advanced basin model in Asia

2. Characteristics of the Citarum River Basin

Recently, severe floods and drought, caused by the global climate change, have occurred in various places around the world. In Asia, in monsoon regions having clear rainy and dry seasons, the water cycle will be accelerated as global warming proceeds, resulting in more intense rainfall and long-term drought. In Indonesia, an increase of food production is needed to accommo-

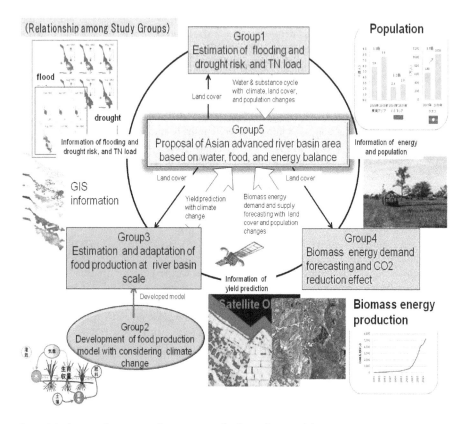

Figure 1. Study groups for proposing the environmentally advanced basin model in Asia

date the growing population. Floods and droughts affect agricultural production directly and indirectly through soil erosion or changes of carbon and nutrient dynamics in the soil. It is difficult to develop more agricultural land in Java, because it is a highly populated island, and thus an irrigation system is necessary to stabilize and increase the production of rice.

The Citarum River is the largest river in West Java. It is 350 km long and has a catchment area of 6000 km^2 (Fig. 2). The annual mean precipitation varies from 1800 to 2800 mm/year [1]. In this basin, 70% of the annual precipitation falls in the rainy season, from November to March. Bandung city is located on the upper part of the river, and there are three large dams on the river (Saguling dam upstream, Cirata dam in the middle of the river and Jatiluhur dam downstream). The Citarum River is the most important river in West Java, supplying water for Bandung and Jakarta. Approximately 80% of the domestic water in Jakarta is withdrawn from the Citarum River downstream of Jatiluhur dam [2]. In the future, climate change may have an undesirable impact on water availability in this basin. Therefore, evaluation of available water resources is quite necessary to manage water resources effectively.

Cihea irrigation area

Figure 2. Citarum River Basin

Classify	(%)
Paddy	35
Plantation	25
Forest	23
Urban, bilt-up	12
Water	5

Table 1. Landuse

Table 1 shows land use in the Citarum basin. The main land uses are paddies (35%), plantations (25%), forest (23%) and urban build-up (12%). The Citarum basin is famous as the region's 'breadbasket' because of its rice production. In this region, the annual agricultural production is mainly paddy rice double production in the wet season (Dec. – Mar. and Apr. – Jul.) without use of an irrigation system. Paddy rice could be planted also in the dry season if enough water was available from rivers or water storage facilities. The Indonesian government constructed the three big dams (Sagling, Cirata and Jatiluhur) for the purposes of electricity, flood control and water use. As a result, in the downstream area, stable irrigation water is supplied year round. Above the dams, the available water fluctuates annually, and there is not enough water for irrigation during the dry season. This drives farmers to cultivate forest into upland fields to produce money crops, which leads to deforestation, soil erosion and nutrient runoff. This is a big factor in the serious sedimentation and eutrophication in the reservoirs [3,4]. To resolve this problem, an effective strategy is needed to alleviate the farmers' poverty, preserve the water quality stored in reservoirs and reduce the runoff from farmlands. The Indonesian government implemented a compensatory system to pay farmers for preserving forest, but the system has not been effective because of the low payment [5]. Furthermore, in the long-term view, compensation will not strengthen farmers' productivity and it will not be effective for dealing with the anticipated population growth in the future. The farmers' poverty must be alleviated in a way that activates and strengthen the existing production system.

3. Current status of agricultural life in the Citarum River Basin

We developed a questionnaire targeting 150 households of farmers randomly sampled from 12,447 in total in the Cihea irrigation area, which is in the midstream area of the Citarum River and in which the farming system practiced is typical for this region, to collect information on agricultural life there. The water resource is the Cisokan River, a branch of the Citarum River, and the total irrigation area is 5,484 ha (Fig. 2). The climate is moderate, and paddy rice production isn't limited by the temperature condition during any part of the year. The annual mean rainfall is over 2,000 mm, and 70% of it falls in the rainy season. The Cisokan River does not have any large water storage capacity in the upstream area, and water intake depends on the natural river discharge. This results in a big seasonal fluctuation of available water from the river, and the irrigation water is not sufficient to grow rice in the dry season. As a typical crop calendar in a year, paddy rice is planted twice in the rainy season (Dec. ~ Mar. and Apr. ~ Jul.) and vegetables such as soy beans and chilies are cultivated extensively in the dry season.

The questions we asked the farmers were designed to gather information about 1) the profiles of respondents, 2) respondents' satisfaction with the rice yield and their income from it, and 3) side jobs respondents performed other than rice production.

1.   Profiles

The average number of household members was 4.3. Figure 3 shows the distribution of harvest area (HA) of each household. HA was defined as follows:

HA = CA (for landowners)

CA×0.5 (for peasants)

where CA is the actual cultivating area. As shown in Fig. 3, the average CA is 0.26 ha, and 98% of all households are small farmers (less than 1 ha).

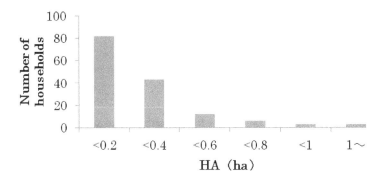

Figure 3. Distribution of harvest area (HA)

2. Satisfaction with rice yield and income

Table 2 shows the respondents' satisfaction with rice yield and income. The values in the table mean the proportion of respondents who answered 'Yes' to each question. We found that most households are not satisfied with rice yield and the subsequent income. On the other hand, it was implied that the poverty level experienced was not so serious as to worry about how to live their lives, because respondents (= head of family) were modestly satisfied with their income and they indicated they would not move to another place if they could get a higher income.

Question	Rate of 'Yes'(%)
Do you want to increase rice yield?	100
Is your income stable every year?	1
Your income is enough to satisfy your family?	18
Your income is enough to satisfy yourself?	58
If you get the higher income, can you move to another place?	2

Table 2. Satisfaction with rice yield and income

3. Side jobs other than rice production

In all, 64% of respondents (95 households) have side job other than rice production. In most cases, they worked at farming operations such as transplanting and harvesting in others' fields or ran small shops that deal with sundry articles. More than half of households with side job get higher income from rice production than from the side job (Fig. 4).

Figure 5 shows the relationship between the number of family members (NF) and the annual rice production in dry weight (P). The line shows the amount of self-consumption (C) related to NF. P and C are estimated as follows:

P = HA ×3.9 (t/ha)×2 (harvests/yr)

C= NF ×100 (kg/capita/yr).

Regardless of NF, many households were plotted around the self-consumption line, and 9 households were plotted under the line, though P varies widely. Considering the rice production cost (=5.0×10⁶Rp./ha for 1 cultivation), annual income (INC) by the surplus rice (S; S=P-C) selling was estimated as follows:

INC=S×1.5×(3.0×10³Rp./kg)×1,000–HA×(5.0×10⁶Rp./ha)×2(harvests/yr)

According to estimated income (Fig. 6), more than 70% of all households had a lower income than the average disposable income of a farmer household in Indonesia (5.8×10⁶Rp., 2008). Furthermore, the fact that some households without side jobs cannot get any income by selling surplus rice indicates that job opportunities are scarce in this area.

The questionnaire results indicated the following about the farmers' consciousness and economic situation. First, the income from selling surplus rice (excluding the farmer's consumption) is not enough to alleviate the farmer's poverty, and the chance of a side job that can compensate for the low income is also not enough to alleviate the farmer's poverty. Second, farmers do not want to move to other places, because they like where they currently live. Therefore, we confirm the necessity of countermeasures using the present paddy rice production system.

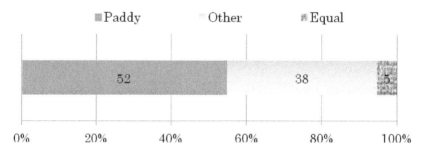

Figure 4. Main income source of households which have side job.

4. Water resources and nitrogen load

4.1. Rainfall -Runoff model

In the Citarum River Basin, the main water user is the agricultural sector [6]. From 1990 to 2008, the average available water from the Citarum River was 7.65×10⁹ m³/year, and the agricultural sector used more than 70% (5.52×10⁹ m³/year) of the total. Using this amount of water, farmers can cultivate irrigated paddies 2.1 times per year on average. For food production, especially in paddy fields, much water is needed, and that is strongly affected by natural

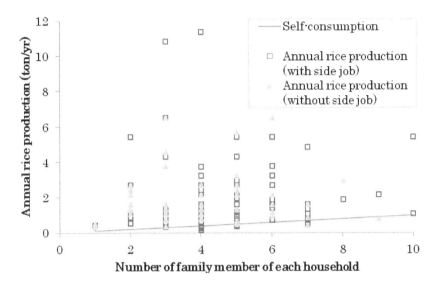

Figure 5. Relationship between number of family member and annual rice production.

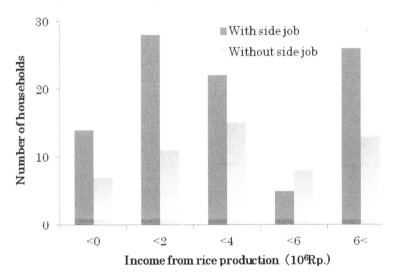

Figure 6. Histogram of income of each household from rice production.

weather conditions. Therefore, to evaluate the future potential of irrigation supply, we developed a distributed water cycling model and used it to analyze the water balance in the Citarum River Basin. For the rainfall-runoff analysis, we used TOPMODEL, a distributed type model. A distributed model can include spatial distribution of topography, land use and soil characteristics. Therefore, this type of model is widely used for hydrological characteristics analysis, water management, water quality analysis and forecasting.

TOPMODEL was proposed by Beven and Kirkdy [7] and is based on the contributing area concept in hill slope hydrology. This model is based on the exponential transmissivity assumption that leads to a topographic index ln($a/To/tanb$). *To* is the lateral transmissivity under saturated conditions, *a* is the upstream catchment area draining across a unit length and *b* is the local gradient of a ground surface. TOPMODEL consists of three soil layers, the root zone, unsaturated zone and saturated zone (Fig.7). Water content (WC) of the root zone and unsaturated zone is calculated by distributed parameters, and WC of the saturated zone is normally calculated by lumped parameters. However, in this study, WC of the saturated zone is also calculated by distributed parameters. Figure 7 shows the TOPMODEL structure. TOPMODEL needs only 3 parameters, so this model is easy to link with GIS data (for details, see [8,9]). A dam operation model was combined with TOPMODEL to evaluate the validation of water storage in the reservoir.

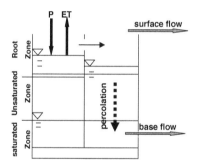

Figure 7. TOPMODEL structure.

4.2. Evaluation of available water resources in the future

For forecasting, we used the Model for Interdisciplinary Research on Climate (MIROC5) with the spatial resolution of 1 degree, which is a general circulation model (GCM), and the generated data was used after bias correction. Figure 8 shows the annual mean air temperature and rainfall amount from 2006 until 2100 of MIROC5(rcp8.5) output at Bandung city. The model predicts that the air temperature will increase gradually more than 3 degrees during approximately 100 years. The trend of annual rainfall will not change significantly. For example, the 10-year average rainfalls are 2193 mm/year (1996 to 2005), 2170 mm/year (2021-2030) and 2258 mm/year (2046-2055).

Figure 9 shows decadal rainfall and calculated river discharge at the Cirata dam station in the cases of (a) 1996-2005 and (b) 2046-2055. In 2046-2055, rainfall intensity became low, with rain falling more equally throughout the year and with middle amounts of rainfall continuing for longer times compared to 1996-2005. The river discharge alters a great deal in response

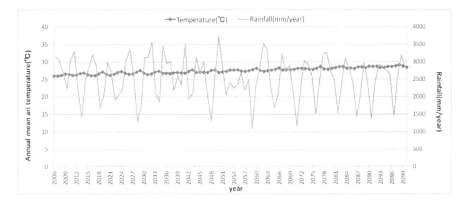

Figure 8. Annual mean air temperature and rainfall amount from 2006 until 2100 of MIROC5 (rcp8.5).

to such rainfall changes. For example, the drought period during which the river discharge was less than100 m³/s in 1996-2005 was 632 days/10 years, but in 2046-2055, it increased to 881days/10years. This increase in the drought period was also affected significantly by evapotranspiration. In this calculation, potential evapotranspiration was calculated from air temperature data by using the Thornthwait method. The runoff ratio, which is the ratio of total river discharge per total rainfall amount, decreased from 0.65 in 1996-2005 to 0.59 in 2046-2055. In the case of flooding (more than 500 m³/s), the frequency increased from 2 times in 1996-2005 to 5 times in 2046-2055. These results show that water will become difficult to obtain in the future, and water scarcity and competition among the water users will be-come severe.

4.3. Nitrogen loads from point sources

High-quality fresh water is limited in quantity, and there is a need for comprehensive water management. Therefore, control of water pollution has become important in many developing countries. In the case of the Citarum River, domestic water for Jakarta and Bandung city has problem of quality and quantity, so water quality assessment is quite important [10]. Although nitrogen is essential for living organisms as an important constituent of proteins, excess enriched nutrients may cause eutrophication of water bodies, especially in lakes or reservoirs. Figure 10 shows the observed total nitrogen (T-N) concentration at Nanjung station, which is located a little bit upstream from Saguling reservoir. The T-N value sometime exceeds 10 mg/L. At Nanjung station, the measurement is affected by waste water from Bandung city and drained water from upstream agricultural fields that flow into the Citarum River.

In the Citarum River, hydrological and water quality data were very limited because of the low frequency of measurements and the fact that data are only observed at a few main stream points. Therefore, it is difficult to evaluate the distribution of the nitrogen pollu-tion load, especially from non-point sources. However, nitrogen loads from point sources such as human and livestock origin can be estimated roughly. Food and Agriculture

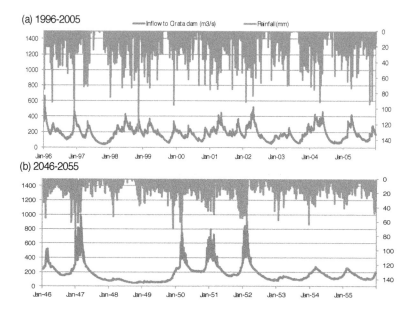

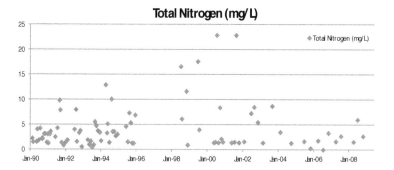

Figure 9. Decadal rainfall and calculated river discharge in 2000 and 2050.

Figure 10. Observed total nitrogen concentration from 1990-2008.

Organization, FAO [11] spatial livestock density data (5-km resolution) and pollution load factors data from Kunimatsu and Muraoka [12] were used for calculation. Figure 11 shows spatial distribution of nitrogen loads from each source in the Citarum basin. Bandung is a highly populated city, so the pollution load is rather large compared to that of a rural area. Table 3 shows annual total nitrogen loads from each source. Humans (54.9%) and chickens (33.8%) were estimated to be the main sources of the nitrogen load in the Citarum

River. In many developing counties, waste water from urban build-up is directly drained into river without treatments. In rural areas, wastes from livestock or human livings are stored in agricultural fields or other soil layers. Such wastes decompose slowly. There-fore, for pollution control or management, a nutrient dynamic model in the soil or water bodies should be developed and applied.

Figure 11. Distribution of nitrogen loads in Citarum river basin.

	Load(kg/year)	rate(%)
Pig	1,081,605	1.7
Cattle	2,581,953	4.1
Chicken	21,461,371	33.8
Goat	3,511,724	5.5
Human	34,927,088	54.9

Table 3. Nitrogen loads frompoint sources

5. Estimation of rice yield from remotely sensed data

5.1. Remote sensing for agriculture

Environmental conservation and food production are critical issues that people in every country must make their best efforts to solve. Remote sensing agricultural research, especially related to rice production and rice field management, is very important for Asian countries, because rice is the staple food for the people and, on the other side, Asian agriculture frequently suffers from heavy losses caused by meteorological events. Considering these matters, it is a good idea to develop an efficient rice cultivation support system based on a concept of precision agriculture that can effectively increase rice production and also realize environmental conservation.

As shown in Table 4, remote sensing using satellites has been tried in a variety of fields and now is considered to be an excellent method for agriculture crop identification [13-15].One of the valuable outcomes to be expected from the remote sensing research for agriculture is to create crop yield models through the assessment of various crop growth variables [16]. For this assessment, it is very important to clarify the quantitative relationship between the remote sensing data and the variables related to crop growth [17]. The remote sensing data to be obtained from crops represents the growing conditions of the crops, which allows the estimation of crop yield [18].

In previous studies of the agricultural remote sensing research in Indonesia, the yield of rice was estimated by the reflectance data of Landsat ETM+ data and MODIS data [19,20]. It was reported that the Ratio Vegetation Index (RVI) was the best vegetation index for both early detection of water deficiency and also distinguishing different crop conditions such as healthy and water stressed crops [21].

Classification	Item
Land	Land use, discrimination of crops, crop acreage, barren farmland, denudation/landslides, melting of snow, floods, damage due to disaster
Crops	Growth conditions, yield estimation, contents and quality, estimating harvest season, outbreaks of agricultural pests, assessing disaster/damage
Soil/ environment	Organic matter content, moisture characteristic, gravel distribution, soil classification, temperature distribution, evapotranspiration, suitability of land
Comprehensive application	Land productivity, land improvement, agricultural field management, growth management, farming plans, preparedness for disaster/damage

Table 4. Uses of remote sensing

5.2. Methodology

In this study, to assess the feasibility of the estimation of rice yield using remotely sensed data, we investigated the relationship between annual rice production from the agricultural statistical data and cumulative Leaf Area Index (LAI) derived from MODIS LAI 8 days composite data in West Java, Indonesia. The study was conducted in 5 sub-districts (Kecamatan Bojongpicung, Ciranjang, Karangtengah, Sukaluyu and Mande) located in the northeast area of Kabupaten Cinajur, West Java, Indonesia (longitude 106°21'E-107° 22'E, latitude 6°42S-7°25'S) (Fig. 12).

The southern part of the mainland West Java has mountains with more than 1,500 m height, although the mountainous area is only 9.5% of the total West Java. The central part is a slope of the mountains with a height from 10 to 1,500 m (36.48%), and the northern part is a large plain with a height from 0 to 10 m (54.03%). Regarding the farming in West Java, 22.89% is mixed farming fields, 20.27% is rice fields and 17.41% is estate crops fields. Only 15.93% of the total area of West Java is forested.

West Java is a major province for agriculture production in Indonesia, but one problem is that the total amount of actually produced crops is less than that of the estimation made by the government. In 2012, the target production amount of rice set by West Java is 12.5 million ton, which is much different from the estimation. Much room remains for improvement of the production amount.

long. 106°21'E - 107°22'E
lat.   6°42'S -   7°25'S

Figure 12. Study site.

In this study, the following data were used for analysis:

1. MODIS/Terra+Aqua Leaf Area Index 8-day L4 Global 1km SIN Grid V005 was collected from October 2001 to September 2008 (276 sets of composite data),

2. SPOT5 satellite (HRG-X) data was acquired on February 20 and July 10, 2011,

3. TERRA satellite (ASTER) data was acquired on May 29,

4. Administrative boundary Geographic Information Systems (GIS) data,

5. Agricultural statistics data from 1996 to 2008 published by Badan Pusat Statistik (BPS).

Figure 13 shows the procedure for analysis. First, the LAI, the SPOT data and the ASTER data were rectified using the administrative boundary GIS data by the nearest neighbor resampling algorithm with use of the selected ground control points. Second, a supervised classification was applied to these rectified images to distinguish the paddy fields, and the mask file of paddy fields was created. Data on cumulative LAI values of paddy fields from October 2001 to September 2008 were calculated, and the data was added to the GIS administrative boundary data. Finally, the tabulate area analysis was executed using the LAI data and the agricultural statistic data to analyze the seasonal trend of LAI and the relationship between the annual rice production and LAI.

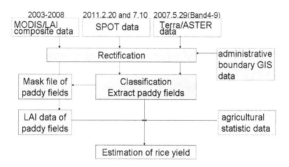

Figure 13. Procedure of image analysis.

5.3. Results and discussion

The seasonal trend of averaged LAI in all sub-districts (Kecamatan) is shown in Fig. 14. There are three minimum values in a year, which are in March, July and December. These seasons correspond to the harvesting season of the rice crop around the test site. In some areas of the test site the vegetation crops are cultivated from August to November, during which period the irrigation water supply is insufficient because of the dry season. This is why the shape of the LAI peak is not clear during this period.

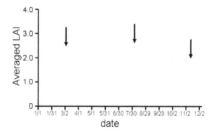

Figure 14. Seasonal trend of averaged LAI of around the test site.

Table 5 shows the relationship between the cumulative LAI of each month and the annual rice production. The result shows significant positive correlation between the annual rice production and the cumulative LAI of each month except for February and December. The correlation coefficients in January, May and September, the months before the harvesting season, respectively, are relatively high against other months. These months are the 30th day or 40th day after transplanting the rice seedlings. The situation in these months is that paddy fields are covered with green crops, and so the water surface of the paddy field is closed.

sub-district	Jan	Feb	Mar	Apr	May	Jun	Jul	Aug	Sep	Oct	Nov	Dec
All	0.558 **	0.353	0.372 *	0.459 *	0.603**	0.657 **	0.529 **	0.671 **	0.668 **	0.604 **	0.569 **	0.217
Bojongpicung	-0.578	0.106	-0.831 *	-0.074	0.380	0.516	-0.266	0.040	0.192	-0.683	0.201	-0.570
Ciranjang	0.120	0.115	-0.648	-0.616	0.542	0.396	-0.398	-0.164	0.488	-0.232	-0.625	-0.740
Karangtengah	0.396	0.023	-0.916 *	-0.141	0.755	-0.031	-0.444	0.261	0.587	0.648	-0.485	-0.878 *
Sukaluyu	0.133	-0.196	-0.887 *	-0.114	0.533	-0.135	-0.378	-0.141	0.268	0.237	-0.758	-0.737
Mande	-0.178	0.169	-0.660	-0.015	0.580	-0.412	-0.027	0.578	0.778	0.294	0.048	-0.805

Table 5. Relationship between cumulative LAI of each month and the annual rice production

On the basis of data from 1996 to 2008, we investigated the relationship between the actual planted acreage and the annual production amount, and we found significant positive correlation in every sub-district.(n=13, p<0.05) (Table 6). Based on these results, the significant positive relationship between the cumulative LAI and the annual production in these months was found, because we could estimate the actual planted acreage from the cumulative LAI data.

All sub-districts	Bojong picung	Ciranjang	Karangtengah	Sukaluyu	Mande
0.781**	0.683*	0.637*	0.854*	0.775**	0.970**

Table 6. Correlation coefficient of between actual planted acreage and the annual rice production

The relationship between the cumulative LAI of January, May and September and the annual rice production is shown in Fig.15. There is a positive correlation between the cumulative LAI of all sub-districts and the annual rice production (r=0.664, p<0.01). Moreover, the correlation coefficient gets higher when limited to three sub-districts where the irrigation ratio is more than 80% (r=0.866, p<0.01).

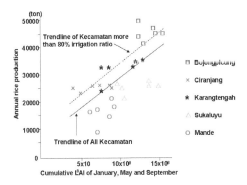

Figure 15. Relationship between the cumulative LAI of January, May and September and the annual rice production.

It was possible to estimate the annual rice production of 2008 using the cumulative LAI of January, May and September from 2003 to 2007. This study indicates that the cumulative LAI of remotely sensed data is applicable to the estimation of rice production amount over a wide area, and the creation of each estimation equation for the irrigated paddy fields and the rain-fed paddy fields will contribute to the improvement of the estimation accuracy of the annual rice production.

6. Toward the development of a simulation model to evaluate geographical distributions of rice growth and yield

6.1. Use of advanced remote sensing for rice production

As the previous section suggests, satellite-based remote sensing is quite useful for obtaining information about crop production. However, the estimation of rice production in the previous section was based on the course relationship between accumulated LAI and rice production, and the relationship lacks crop ecological mechanisms. Accordingly, if the crop ecological processes are incorporated into the relationship, the accuracy of estimation of crop production should improve. Such processes could estimate not only the production but also the cause for the production by inverse calculation. If the inverse calculation can suggest the constraints of production, e.g., soil fertility, water management and cultivar, the suggestions will be quite useful for improving the productivity and developing a strategy for the future. To incorporate crop ecological processes into the relationship between remote sensing data and crop production, we have been developing a simulation model combined with remote sensing (SIMRIW-RS) [22]. This section explains the concept of the simulation model and reports the present status of development.

6.2. The concept of a simulation model combined with remote sensing

The architecture of our simulation model, SIMRIW-RS was derived from SIMRIW-Rainfed, which was developed on the basis of SIMRIW [23] to simulate variation in rice productivity at farmers' paddy fields in Northeast Thailand [24]. Since SIMRIW is widely adapted to simulate rice productivity under a future climate scenario [25,26], its derivative simulation models are also expected to be able to simulate that.

To parameterize for crop growth by remote sensing data, smaller numbers of parameters in the simulation model are better. On the other hand, the simulation model is required to evaluate variation in productivity among farmers' fields. Evaluation of cultivar difference is also required to select optimum cultivars to improve the productivity and to develop a strategy of cultivar breeding for adaptation to climatic change in the future. These requirements indicate that the simulation model should have the minimum number of field and cultivar parameters that evaluate field-to-field and cultivar-to-cultivar variability.

To simulate phenological development of rice, the same functions in SIMRIW are applied. Namely, the phenology is indexed by a Developmental Index (DVI): DVI = 1 for panicle initiation, DVI = 2 for heading and DVI = 3 for maturity [27]. DVI progress is expressed by the accumulation of the daily progress rate of development (DVR), which is expressed by functions of day length and daily average temperature. Although these functions need some cultivar parameters, the parameters can be estimated based on the database of parameters [28] and adjusted by the present cultivation schedule. If phonological development can be distinguished by satellite-based remote sensing, the parameters can also be estimated. However, since the distinguishing is not very accurate yet, the estimation of cultivar parameters for DVI by remote sensing is not under consideration in this study.

Plant growth and yield formation are expressed by four functions (Table 7). Plant growth is driven by nitrogen uptake (Nup), which is expressed by a function with a field parameter for nitrogen supply and a cultivar parameter for nitrogen uptake ability. Leaf area (LA) expansion is a function of Nup and DVI with two cultivar parameters: the LA expansion rate and the maximum LAI. Since the maximum LAI is set as a constant value (= 7), a substantial number of parameters for LA expansion is one. Dry matter (DM) is a product of solar radiation intercepted by LA and a cultivar parameter radiation conversion efficiency, which is varied with Nup per LA. Grain yield is obtained by DM multiplied with a cultivar parameter, the harvest index. Consequently, one field parameter and four cultivar parameters must be determined to simulate rice growth and yield.

Function	Note[1]
(1) Phenological development	Nakagawa and Horie (oxox); D_l: day length; T_{ave}: daily average temperature; cultivar parameters are offered
$DVI = \Sigma\, DVR$	
$DVR = f(D_l, T_{ave})$	
(2) Nitrogen uptake (Nup)	
$\Delta Nup = f(\alpha_1, \beta_1, T_{ave}, DVI)$	α_1: nitrogen supply
	β_1: nitrogen uptake ability
(3) Leaf area (LA) expansion	
$\Delta LA = f(\beta_2, \beta_3, \Delta Nup, DVI)$	β_2: Leaf area expansion rate
	β_3: Maximum LAI (= 7 in this study)
(4) Dry matter (DM) production	
$\Delta DM = f(\beta_4, S_n, Nup, LA, DVI)$	S_n: Solar radiation
	β_4: Radiation conversion efficiency
(5) Yield formation	
$Yield = f(\beta_5, DM)$	β_5: Harvest index

[1] α: field parameter; β: cultivar parameter.

Table 7. List of functions in the simulation model [22]

6.3. Determination and validation of default parameters in the simulation model

To test model performance, we determined default parameters in the simulation model based on a field experiment. The field experiment was conducted in Kyoto using three different types of field. In one of the three types of field, three kinds of fertilizer treatment were tested, and thus five nutritional environments were tested. Six different types of cultivar were planted in each environment (Table 8).

Field (α_1) and cultivar (β_1) parameters for Nup were determined statistically based on the relationship between Nup by plant and accumulated effective average air temperature. The other three cultivar parameters (β_2, β_4 and β_5) were optimized by a simplex method, one of the nonlinear least squares methods, based on the observed data.

The obtained parameters are shown in Table 8. The parameters seem to express the field and cultivar characteristics. Based on the parameters, the simulation model well explained field-to-field and cultivar-to-cultivar variation in Nup, LA, DM and yield (Fig. 16), suggesting that this use of one field parameter and four cultivar parameters was adequate for the purpose.

Field parameter	Kyoto Univ. Expt. field				Conseq. Un-fert. Field	
	Un-fert.	Less fert.	Standard		6 yr conseq.	59 yr conseq.
α_1: N supply	0.0045	0.0049	0.0076		0.0034	0.0019
	Nipponbare	Kasalath	Bei Khe	Takanari	B6144F	Baniasahi
β_1: NUA	0.0043	0.0048	0.0042	0.0054	0.0044	0.0037
β_2: LER	3.3	4.9	5.2	3.1	5.9	5.0
β_4: RUC	0.46	0.66	0.64	0.49	0.69	0.58
β_5: HI	0.43	0.46	0.39	0.48	0.45	0.44

fert.: fertilizer; conseq.: caonsequence; NUA: nitrogen uptake ability; LER: leaf expansion rate; RUC: radiation conversion efficiency; HI: harvest index.

Nipponbare: improved japonica in Japan; Kasalath: traditional indica in India; Bei Khe: traditional indica in Cambodia; Takanari: improved indica in Japan; B6144F: improved indica in Indonesia; Baniasahi: traditional japonica in Japan.

Table 8. List of default field and cultivar parameters in the simulation model. Values were determined based on the field experiment

6.4. Remote sensing technique for optimizing field and cultivar parameters in the simulation model

As mentioned in previous section, the simulation model can do a good job of estimating rice growth when one field parameter and four cultivar parameters in the simulation model can be set appropriately. Remote sensing will be helpful for optimizing these parameters using a simplex method on the regional and global scale. While there are several procedures to optimize the above-mentioned parameters using remote sensing, that shown in Fig. 17 was

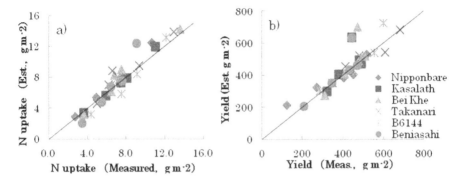

Figure 16. Comparison between measured and estimated nitrogen (N) uptake (a) and yield (b). Six different cultivars tested in the five different nutritional environments.

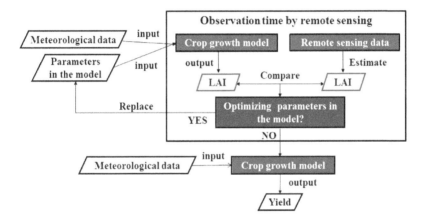

Figure 17. Concept of optimizing field and cultivar parameters in the simulation model using remote sensing data.

selected for this study. This figure shows that these parameters will be optimized by comparison between LAI simulated by the model and the one estimated by remote sensing data. This figure indicates that LAI is a key factor and that the estimation accuracy of LAI obtained from remote sensing data is the most important source of information for optimizing these parameters in the model.

When remote sensing is used to monitor crop LAI, there are two methods to determine the desired information. One is to use the regression expression obtained from the relationship between vegetation index derived from spectral reflectance at the crop canopy level and LAI measured by field experiment, and another is to use the inversion method of the radiative transfer model with spectral reflectance at the crop canopy level. Generally, at the satellite (or airborne) level, it is more difficult to apply the inversion method to monitor crop LAI than the regression expression obtained from the relationship between vegetation indices derived from spectral reflectance and LAI measured by field experiment. Therefore, the method using vegetation indices has been widely used to monitor crop LAI on the regional and global scale. Popular vegetation indices for monitoring crop LAI include the Normalized Difference Vegetation Index (NDVI), the Soil Adjusted Vegetation Index (SAVI) and the Enhanced Vegetation Index (EVI). The most famous index is the NDVI [29]. The value of the NDVI increases with an increase in the amount of crop leaves. Although the NDVI was widely used to estimate crop LAI, it was reported that the value of NDVI saturated when plants at grew closely or when the plant canopy structure was complex and consisted of multiple layers [30]. The EVI was developed to solve this problem of the NDVI [31]. The EVI can be applied to the situation when plants at the observed place have high density or a complicated structure, and this index reduces the adverse effects of environmental factors such as atmospheric conditions and soil background [32]. The SAVI is also an improved version of the NDVI. The SAVI is the index that considers the effect of soil background on the NDVI [33].

The above-mentioned indices have been used for plant monitoring. Generally, the regression expression obtained from the relationship between vegetation indices derived from remote sensing data at the crop canopy level and measured LAI at the ground level have been widely used to monitor plant LAI. However, these regression expressions were not derived based on a radiative transfer process of reflectance dependent on crop growth. Namely, these regression expressions have the possibility of changing from year to year and place to place. Here, a new vegetation index for monitoring plant LAI developed by the authors is described. This index was derived from the radiative transfer process of reflectance that is dependent on crop growth. The new vegetation index is based on the results reported in Oki et al. [34] and described both the accuracies of mixel (mixed pixel) decomposition when spectral reflectance at visible wavelength bands was used and when spectral reflectance from visible to near-infrared wavelength bands was used. In general, the reflectance value from a mixel is expressed as the linear combination of the reflectance values of the target and other objects, as follows:

$$\mathbf{p} = \alpha\mathbf{m} + (1 - \alpha)\mathbf{n}$$

$$\mathbf{p} = \begin{pmatrix} p_1 \\ \vdots \\ p_N \end{pmatrix}, \quad \mathbf{m} = \begin{pmatrix} m_1 \\ \vdots \\ m_N \end{pmatrix}, \quad \mathbf{n} = \begin{pmatrix} n_1 \\ \vdots \\ n_N \end{pmatrix} \tag{1}$$

where p, m and n are vectors of measured spectral reflectance, pure reflectance of the target object and noise, respectively. Here, α is the fraction of the target object area, and N is the number of wavelength bands. To estimate the fraction of the target object area, Oki et al. [34] developed an improved matched filter method that estimates the fraction of the target object area by calculating the degree of similarity of wave profiles of measured spectral reflectance to pure spectral reflectance of a target object. According to Oki et al. [34], using spectral reflectance only in the visible region is more suitable for precise detection of the fraction of the target object area than is using spectral reflectance at the visible and near-infrared wavelength bands. Reflectance at the visible wavelength band is insusceptible to multi-scattering in a canopy, because reflectance and transmittance of this wavelength of leaves are small. Therefore, α calculated using only reflectance at visible regions reflects information about the crop canopy surface. On the other hand, reflectance and transmittance at near-infrared wavelength bands of leaves are larger than those at visible wavelength bands. Therefore, the accuracy of α degrades because the influence of multi-scattering in a canopy is included in α' when reflectance at visible and near-infrared wavelength bands is used to calculate α. The influence of multi-scattering becomes larger when a value of crop LAI becomes larger, because multi-scattering mainly depends on the LAI value. Therefore, the difference between fractions derived using reflectance at visible regions and reflectance from visible to near infrared regions indicates the influence of the LAI value, as follows:

$$L_{dif} = \alpha_{vis} - \alpha_{vis+nir}$$
$$= \text{influence of multi scattering at near - infrared regions} \qquad (2)$$
$$\approx \text{influence of LAI}$$

where α_{vis} is the fraction of the target area to the unit ground area, and $\alpha_{vis+nir}$ is the fraction of the target area to the unit ground area including the influence of multi-scattering at near-infrared wavelength bands.

To verify the availability of L_{dif} under several LAI-coverage relationships, we set up two relationships (Fig. 18). This verification was performed using the radiative transfer model known as Forest Light Environmental Simulator (FliES), developed by Kobayashi and Iwabuchi [35]. FliES can simulate the reflectance at the top of the canopy under several conditions. LAI-coverage conditions were set up by changing the parameters such as plant density, leaf area density and plant size in FliES. The LAI used in this figure was measured by a field experiment conducted in the Kyoto University Experimental Field in 2008. Figure 19 is the time-series changes of conventional indices and the new index under two LAI-coverage conditions. The values of these indices range from 0 to 1. This figure indicates that conventional indices are more sensitive to the change of coverage than is L_{dif}. Namely, the new index relates to only LAI under several planting forms. Fig. 20 shows the time-series changes of LAI and these indices. The values of each index in this figure were normalized using minimum and maximum values of each index. Table 9 shows root mean square errors of normalized vegetation indices for normalized LAI. This figure and table indicate that the normalized L_{dif} can directly depict the time-series pattern of the LAI. Hence, we believe our new index is more useful for estimating the LAI than conventional indices. Although L_{dif} is useful to estimate rice LAI, calculating the index value requires pure reflectance of the target object and maximum reflectance at each wavelength band. Generally, pure reflectance of rice may be measured during the heading period, because the coverage and LAI of rice are at maximum during this period. Maximum reflectance at visible wavelength bands may be measured before the transplanting period, and that at near-infrared regions may be measured during the heading period. Therefore, the setting of these reflectances for a target area should be discussed before applying this index in a future study. L_{dif} may be one of the most useful indices to estimate

crop LAI when the way to set these reflectances is determined.

	Normalized vegetation index							
	N-NDVI 1	N-NDVI 2	N-EVI 1	N-EVI 2	N-SAVI 1	N-SAVI 2	N-Ldif 1	N-Ldif 2
RMSE	0.132	0.170	0.074	0.104	0.084	0.114	0.057	0.063

Table 9. Root mean square errors of normalized vegetation index for normalized LAI

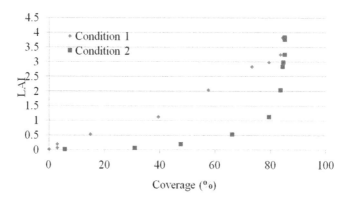

Figure 18. LAI-coverage relationships used to verify the availability of new index.

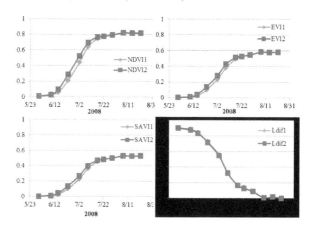

Figure 19. The time-series changes of vegetation indices under two LAI-coverage conditions. The suffixes of each index are corresponding to the suffixes of LAI-coverage conditions.

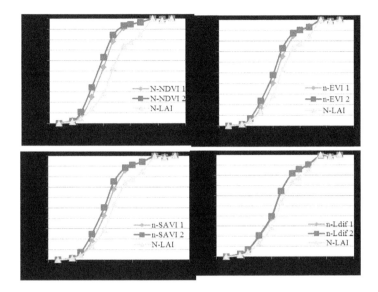

Figure 20. Time-series pattern of vegetation indices and LAI

7. Impact of 2nd generation biofuel development in Indonesia

7.1. 1st generation and 2nd generation biofuels

To reduce the amount of imported fuel, lower the poverty ratio and reduce greenhouse gas (GHG) emissions, the Government of Indonesia is attempting to find alternative renewable energy, particularly in the form of biofuel. However, it is afraid that promotion of 1st generation biofuel produced primarily from food crops may cause deforestation or compete with food [36]. It is increasingly understood that 1st generation biofuels are limited in their ability to achieve targets for oil-product substitution. The cumulative impacts of these concerns have increased the interest in developing biofuels produced from non-food biomass, such as agricultural residues. These "2nd generation biofuels" could avoid many of the concerns facing 1st generation biofuels. Some studies have estimated the potential of agricultural residuals, such as rice straw, for ethanol production [37,38]; however, little is known about the potential economic impact of producing 2nd generation biofuels. The purpose of this part of our study was to estimate the potential production of rice straw for ethanol production and examine the possible economic impact of such production in Indonesia.

7.2. The present situation of biofuel development and production in Indonesia

The Government of Indonesia has released a new energy policy that aims at increasing energy self-sufficiency by publishing a road map for biofuel development as an alternative renewable energy based on the National Energy Policy (Presidential Decree No. 5/2006). This road map incorporates a plan to gradually increase the use of biofuels so that they accounts for 5% of the total energy supply by 2025. Table 10 shows details of the road map for increasing the use of biofuels. Among the biofuels proposed, biodiesel is made from palm oil and jatropha with the goal of having biodiesel account for 10% of total diesel consumption by 2005-2010, 15% by 2011-2015 and 20% by 2016-2025. In the bioethanol sector, which uses sugar cane and cassava as feedstock, it is planned to increase the share of bioethanol of total gasoline consumption to 5% by 2005-2010, 10% by 2011-2015 and 15% by 2016-2025. Based on these goals, it is estimated that 10.22 million kiloliters of biodiesel and 6.28 million kiloliters of bioethanol will be required per year by 2025 (Table 10). It is planned that cassava and sugar cane will be the major feedstock for bioethanol, while oil palm and jatropha will be the major feedstock for biodiesel. However, there is a wide gap between the potential consumption and actual consumption (Fig.21). Indonesia's fuel ethanol production has remained at the zero level since 2010 due to disagreement in market price index formulation between MEMR (The Indonesian Ministry of Energy and Mineral Resources) and ethanol producers [39]. Moreover, it is not easy to increase the production of feedstock of bioethanol without forest degradation. Moon and Shirakawa [36] point out that in the case of 1st generation biofuel, deforestation is inevitable to achieve Indonesia's biofuel provision target, and there will be a negative effect of increasing the net CO2 emission due to massive expansion of land use for biofuel cultivation. On the other hand, there is great potential for ethanol production from agricultural residues in Indonesia [37]. Therefore, it is important to develop the necessary technology to produce bioethanol from agricultural residues, especially from rice straw.

		2005-2010	2011-2015	2016-2025
Biodiesel (Palm oil, Jatropha)	Plan	10% of Biodiesel of total Diesel consumption	15% of Biodiesel of total Diesel consumption	20% of Biodiesel of total Diesel consumption
	Amount	2.41 million kl	4.52 million kl	10.22 million kl
Bioethanol (Sugar cane, Cassava)	Plan	5% of Bioethanol of total Gasoline consumption	10% of Bioethanol of total Gasoline consumption	15% of Bioethanol of total Gasoline consumption
	Amount	1.48 million kl	2.78 million kl	6.28 million kl
Biooil — Bio Kerosene	Plan	Using Bio Kerosene	Using Bio Kerosene	Using Bio Kerosene
	Amount	1.00 million kl	1.80 million kl	4.07 million kl
Pure plantation oil	Plan	Using PPO	Using PPO	Using PPO
	Amount	0.40 million kl	0.74 million kl	1.69 million kl
Total Biofuel	Plan	2% of Biofuel of total Energy consumption	3% of Biofuel of total Energy consumption	5% of Biofuel of total Energy consumption
	Amount	5.29 million kl	9.84 million kl	22.26 million kl

* Source: National Plan on Biofuel, "Development of Alternative Energy in Indonesia".

* PPO: Pure Plantation Oil(oil derived from jatropha)

* kl: Kiloliter

Table 10. Roadmap for increasing use of biofuel in Indonesia

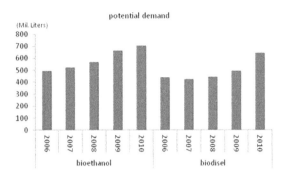

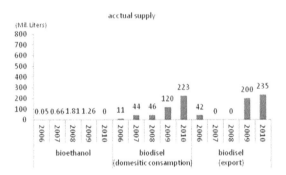

Note: potential consumption is equal to bending rate multiplied by gasoline / diesel oil consumption. Source: [39]

Figure 21. Indonesia biofuel use in transportation sector.

8. Proposal of a poverty alleviation policy for farmers as an example of an environmentally advanced basin model in Asia with a focus on water, food and energy

In this section, we propose a scenario for alleviating farmers' poverty that activates unused resources in the Cihea irrigation area. As unused resources, water resource development and rice straw are assumed.

1.   Water resource development

Figure 22 shows precipitation, river discharge and intake water for irrigation. Each value is the monthly average for 9 years (2001 to 2009). As mentioned above, the Cisokan River does not have any large water storage in the upstream area, and water intake depends on the natural river discharge. This results in a big seasonal fluctuation of available water from the river, and the water supply is not sufficient for irrigation in the dry season.

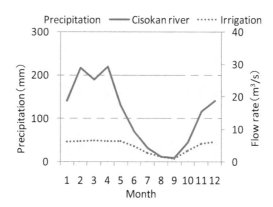

Figure 22. Precipitation, river discharge, and water intake for irrigation (monthly average data from 2001 to 2009).

The daily water demand was calculated as the sum of evapotranspiration (5 mm/day) and infiltration (10 mm/day), and the irrigation water surplus/shortage was estimated by comparing the daily water demand with the intake water and effective rainfall. After estimating the amount of irrigation water, we determined that the water shortage in the dry season and surplus in the rainy season would be 7.1 mm/day and 25.4 mm/day, respectively, and if one-third of the water surplus in rainy season could be stored, enough water would be supplied for rice production in the dry season over the entire irrigation area. Furthermore, since the solar radiation in the dry season is more than that in the rainy season, the yield in the dry season is expected to be more than that in the rainy season.

2.   Economic Impact of Biofuel Development from Rice Straw

In the Citarum River Basin, most rice straw is burned, with a small part of it used for feeding. This unused resource can be utilized as a material for bioethanol. If rice straw left over after harvest is used, it would have a positive economic impact on farmers, both because of selling it and in the form of payment for collection.

We conducted field survey in this area in Jul. 2010 and Mar. 2011. As a result, the average rice grain yield in the rainy season was 5.8 t/ha (fresh weight). The rice straw produced can thus be calculated as follows:

5.8 t/ha×2/3×1.55.8 t/ha

where it is assumed that the dry weight is 2/3 of the fresh weight and the rice straw weight is 1.5 times the dry weight.

As a model case, we estimated the economic impact that can be expected by activating unused resources for a typical household, that is, NF = 4 (parents and two children), HA = 0.26 ha and without side jobs. In this case, the values associated with rice production are estimated as follows:

P = 2.0 t/yr, S = 1.6 t/yr and INC $=4.7\times10^6$Rp/yr

As mentioned above, in the Citarum River Basin, water resource development would enable three annual planting times, and thus the value of INC would increase to 8.0×10^6 Rp/yr.

When rice straw is used as a material of bioethanol, the income from selling and collecting the rice straw can be expected. The economic impact of selling rice straw is estimated using the price of 100 Rp/kg, while it is considered to be 0 to 200 Rp/kg for feeding in Indonesia. The annual income from selling rice straw is thus estimated as follows:

0.26 ha$\times5.8\times10^5$Rp/ha$\times2$ harvest times/yr = 3.0×10^5Rp/yr

This value is equal to 6.3% of INC. To collect rice straw after harvest, the amount of work is assumed to be 3 day/ha by one person (Saga et al. 2008), and payment for collection is assumed to be the same as for harvest, 4.0×10^4 Rp/day. The income of farmers from collection is estimated as follows:

0.26 ha$\times2$ harvest times$\times3$ days/ha$\times(4.0\times10^4$Rp/day$)$ = 6.2×10^4Rp/yr

These estimates are summarized in Table 2. Planting three times per year by the development of water resource and making bioethanol from rice straw would enable farmers to increase their income by 3.7×10^6 Rp/yr, up to a 1.8-fold increase.

9. Conclusions

In this study, we carried out the following:

1. Outline the current status of agricultural life in the Citarum River Basin in Indonesia.
2. Quantify the uneven distribution of water resources and change in nitrogen load at the basin level, and assess the varietal diversity against the drought risk.
3. Assess the paddy rice yields with remotely sensed data.
4. Propose a simulation model combined with remote sensing to evaluate geographical distributions of rice growth and yield.
5. Outline a strategy for the sustainable use of biomass energy at the regional and basin levels.

Utilizing the preceding five items, we proposed an environmentally advanced basin model in the Cihea irrigation area. In particular, we proposed a poverty alleviation policy for farmers as an example of an environmentally advanced basin model in Asia with a focus on water, food and energy.

Rice is the most important crop in Indonesia, and large quantities of rice residues in the form of straw are available. Our findings show that the production of biofuel will contribute to poverty alleviation. However, rice straw is also used for organic fertilizer. It is necessary to determine what percentage of rice straw could be used for bioethanol from the viewpoint of a sustainable rice yield. The cost of ethanol production from rice straw is still higher than the cost of gasoline; however, the production of bioethanol from agricultural residuals could create jobs and contribute positively to the energy security issue. We need to consider the cost-benefit of biofuel production not only on the plant level but also on the social level.

Our study results confirm the necessity of taking measures to improve the present paddy rice production system. It is estimated that the three-times production per year made possible by the development of the water resource and making bioethanol from rice straw would enable farmers to increase their income up to 1.8-fold.

Regarding future plans, it is necessary to form a cooperative relationship among local government, researchers and agricultural stakeholders and examine more concretely the possibility of conducting the suggested measures in real society to alleviate the poverty of farmers.

Acknowledgements

This research was supported by environment research & technology development fund (E-1104: Development and Practice of Advanced Basin Model in Asia : Toward Adaptation of Climate Changes (FY2011–FY2013), Ministry of the Environment, Japan). We would like to express our gratitude for its financial aid.

Author details

Kazuo Oki[1], Keigo Noda[1], Koshi Yoshida[2], Issaku Azechi[2], Masayasu Maki[3], Koki Homma[3], Chiharu Hongo[4] and Hiroaki Shirakawa[5]

1 The University of Tokyo, Japan

2 Ibaraki University, Japan

3 Kyoto University, Japan

4 Chiba University, Japan

5 Nagoya University, Japan

References

[1] BPSDA(Balai Pengelolaan Sumber Daya Air)((2006). Status Report Balai PSDA WSCitarum (in Indonesian)

[2] Loebis, J, & Syamman, P. (1993). Reservoir operation conflict in Citarum river basin management, IAHS Pub. , 213, 455-459.

[3] ADB(Asian Development Bank) ((2004). Technical assistance to the republic of Indonesia for preparing the integrated Citarum water resources management projectADB report TAR: INO 37049, Directorate General for Water Resources, Ministry of Public Works, Jakarta.

[4] Basyar, P. A, & Harto, B. (2006). Spatial modeling of sediment transport over the upper Citarum catchment, PROC. ITB Eng. Science, 38(1), 11-27.

[5] Pirard, R, & Bille, R. (2010). Payments for Environmental Services (PES): a reality check (stories from Indonesi), Analyses IDDRI.(3)

[6] BBWSC(Balai Besar Wilayah Sungai Citarum) (2011): Profil BBWSC (in Indonesian).

[7] Beven, K. J, & Kirkby, M. J. (1979). A physically based, variable contributing area model of hydrology., Hydrological science bulletin. 24(1), 43-69.

[8] Ao, T, Ishihira, H, & Takeuchi, K. (1999). Study of Distributed Runoff Simulation Model Based On Block Type Topmodel and Muskingum-Cunge Method, Annu. J. of Hydr. Eng., , 43, 7-12.

[9] NawarathnaNMNS B., Ao, T. Q., K azama, S., Sawamoto, M. and Takeuchi, K. ((2001). Influence of Human Activity on the BTOPMC Model Runoff Simulations in Large-scale Watersheds., XXIX IAHR congress proceedings, theme a, , 93-99.

[10] Fares, Y. R. (2003). Water resources management in tropical river catchments.,journal of Environmental Hydrology. no. Papers 14, , 11, 1-11.

[11] FAO : Livestock density (GEONETWORK)(2005). http://www.fao.org/AG/againfo/resources/en/glw/ GLW_dens.html.

[12] Kunimatsu, T, & Muraoka, K. (1989). Model Analysis of River Pollution (in Japanese), Gihoudo Press(Tokyo).

[13] Bouvet, A. Le Toan, T. and Lam-Dao, N. ((2009). Monitoring of the rice cropping system in the Mekong delta using ENVISAT/ASAR dual polarization data. IEEE Trans. Geoscience and Remote Sensing, 47, , 517-526.

[14] Pan, X. Z, Uchida, S, Liang, Y, Hirano, A, & Sun, B. (2010). Discriminating different landuse types by using multitemporal NDXI in a rice planting area. International Journal of Remote Sensing, , 31, 585-596.

[15] Niel, T. G. V, & Mcvicar, T. R. (2003). A Simple method to improve field-level rice identification: toward operational monitoring with satellite remote sensing. Australian Journal of Experimental Agriculture, , 43, 379-387.

[16] Ahlrichs, J. S, & Bauer, M. E. (1983). Relation of agronomic and multispectral reflectance characteristics of spring wheat canopies. Agronomy Journal, , 75, 987-993.

[17] Patel, N. K, Singh, T. P, Sahai, B, & Patel, M. S. (1985). Spectral response of rice crop and its relation to yield and yield attributes. International Journal of Remote Sensing, , 6, 657-664.

[18] Barnett, T. L, & Thompson, D. R. (1982). The use of large-area spectral data in wheat yield estimation. Remote Sensing of Environment, , 12, 509-518.

[19] Nuarsa, I. W, Nishio, F, & Hongo, C. (2011). Relationship between Rice Spectral and Rice Yield Using Modis Data, Journal of Agriculture Science, Canada, , 3(2)

[20] Nuarsa, I. W, Nishio, F, & Hongo, C. (2012). Rice Yield Estimation Using Landsat ETM+ Data and Field Observation, Journal of Agriculture Science, Canada, , 4(3)

[21] Nuarsa, I. W, Nishio, F, & Hongo, C. (2011). Spectral characteristic comparison of rice plants under healthy and water deficient conditions using Landsat RTM+ data, Journal of the Japan Society of Photogrammetry and Remote Sensing, , 50(2)

[22] Homma, K, Maki, M, & Hirooka, Y. (2012). Toward to develop a simulation model to evaluate geographical distributions of rice growth and yield. Jpn J. Crop Sci. 81 (Extra (1)

[23] Horie, T. (1987). A model for evaluating climatic productivity and water balance of irrigated rice and its application to Southeast Asia. South Asia Studies , 25, 62-74.

[24] Homma, K, & Horie, T. (2009). The present situation and the future improvement of fertilizer applications by farmers in rainfed rice culture in Northeast Thailand. In: L.R. Elsworth, W.O. Paley (Eds.) Fertilizers: Properties, Applications and Effects. Nova Science Publishers, N.Y. , 147-180.

[25] Horie, T, Nakagawa, H, Centeno, H. G. S, & Kropff, M. (1995). The rice crop simulation model SIMRIW and its testing. In Modelling the Impact of Climate Change on Rice in Asia (Matthews, R. B. et al. eds.), CAB International, Oxon, UK. , 51-66.

[26] Tanaka, K, Kiura, T, Sugimura, M, Ninomiya, S, & Mizoguchi, M. (2011). Tool for predicting the possibility of rice cultivation using SIMRIW. Agric. Inform. Res. , 20, 1-12.

[27] Nakagawa, H, & Horie, T. (1995). Modeling and prediction of developmental process in rice. II. A model for simulating panicle development based on daily photoperiod and temperature. Jpn. J. Crop Sci. , 64, 33-42.

[28] Nakagawa, H, & Horie, T. (1997). Phenology determination in rice. In: Breeding Strategies for rainfed lowland rice in drought-prone environments (Fukai, S. et al. eds.). Australian Centre for International Agricultural Research, Canberra. , 81-88.

[29] Rouse, J. W, Haas, R. H, Schell, J. A, Deering, D. W, & Harlan, J. C. (1974). Monitoring the vernal advancement and retrogradation of natural vegetation. NASA/GSFC Final Report, , 1-137.

[30] Baret, F, & Guyot, G. (1991). Potentials and limits of vegetation indices for LAI and APAR assessment. Remote Sensing of Environment , 35, 161-173.

[31] Liu, H. Q, & Huete, A. R. (1995). A feedback based modification of the NDV I to minimize canopy background and atmospheric noise. IEEE Transactions on Geoscience and Remote Sensing , 33, 457-465.

[32] Huete, A. R, & Justice, C. (1999). MODIS vegetation index (MOD13) algorithm theoretical basis document. NASA EOS MODIS Doc., Version 3.0.

[33] Huete, A. R. vegetation index (SAVI). Remote Sensing of Environment , 25, 295-309.

[34] Oki, K, & Oguma, H. (2002). Estimation of the canopy coverage in specific forest using remotely sensed data- Estimation of Alder trees in the Kushiro Mire-. Journal of the Remote Sensing Society of Japan , 22, 510-516.

[35] Kobayashi, H, & Iwabuchi, H. D atmosphere and 3-D canopy radiative transfer model for canopy reflectance, light environment, and photosynthesis simulation in a heterogeneous landscape. Remote Sensing of Environment , 112, 173-185.

[36] Moon, D, & Shirakawa, H. (2012). Increase in biofuel use and corresponding changes in land use in Indonesia, Journal of Environmental Information Science, , 40(5), 69-78.

[37] Yano, S. (2010). Study on Bioethanol Production From Agricultural Residuals in Asia, http://repository.dl.itc.u-tokyo.ac.jp/dspace/bitstream/2261/37429/1/39_077098.pdf

[38] Diep, N. Q, Fujimoto, S, Minowa, T, Sakanishi, K, & Nakagoshi, N. (2012). Estimaion of the potential of rice straw for ethanol production and the optimum facility size for different regions in Vietnam, Applied Energy, , 93, 205-211.

[39] USDA foreign Agricultural ServiceIndonesia Biofuel Annual ((2011). http://gain.fas.usda.gov/Recent%20GAIN%20Publications/Biofuels%20Annual_Jakarta_Indonesia_8-19-2011.pdf

Chapter 3

Molecular Mechanisms Controlling Dormancy and Germination in Barley

Santosh Kumar, Arvind H. Hirani,
Muhammad Asif and Aakash Goyal

Additional information is available at the end of the chapter

1. Introduction

1.1. History of barley

Barley (*Hordeum vulgare* L.) is amongst the oldest crops within cereals. Archaeological remains of this crop have been discovered at different locations in the Fertile Crescent (Zohary & Hopf, 1993) indicating that barley is being cultivated since 8,000 BC. The wild relatives of barley were recognized as *Hordeum spontaneum* C. Koch. However, in the recent literature of taxonomy, *H. spontaneum* C. Koch, *H. vulgare* L., as well as *H. agriocrithon* Åberg, are believed to be the subspecies of *H. vulgare* (Bothmer & Jacobsen, 1985). Studies with molecular markers have confirmed that barley was brought into cultivation in the Isreal-Jordan area but barley diversification occurred in Indo-Himalayan regions (Badr et al., 2000).

1.2. Importance of barley in Canada

Barley, a gladiator's food in Athens and the only crop to be used as a form of money in early Sumerian and Babylonian cultures, is the fourth largest cultivated crop in the world after wheat, rice and maize. Barley is one of the most fundamental plants in human nutrition and it is one of the most widely cultivated cereal grown in various climatic regions of world; starting from sub-Arctic to subtropical (Zohary & Hopf, 1993). Depending on the physical arrangement of the kernels on the plant, it is categorized into two different types as six-row and two-row barley. Based on the presence or absence of covering on the kernels, it is also classified as hulled or hull-less.

In Canada, it was first cultivated in Port Royal in 1606. Today, Canada is the 4th biggest barley producer after the European Union, Russia and Ukraine (Taylor et al., 2009). Most farmers grow barley for sale as malting barley. If the grain does not meet malting quality, it is sold as feed barley. Malting quality is somewhat subjective and depends upon the supply of good malting barley and its price. In the past couple of years, barley crops have suffered great loss in yield and quality due to lower germination potential and water sensitivity (Statistics Canada, 2007). Despite significant losses in barley production and yield in the year 2006-2007 (9.5 million tonnes (Mt)), the total production of barley increased (11.8 Mt) in 2007-2008 due to larger cropping area at the expense of wheat acerage (Statistics Canada, 2007).

Total barley production decreased by 10% and the harvested area by 1.5% in 2009 compared to 2008. Domestic use has increased by 4% due to a decline in corn imports. Total exports have increased by 12.5% in 2009 after a drastic decline of 47% in 2008 from the previous year (USDA Report, 2009). Average price for malt barley has gone down significantly from $208 to $179 per tonne (Agric. & Agri-Food Canada, 2009).

1.3. Challenges related to barley production

Malting quality characteristics (beta-glucan content, protein breakdown, fermentability, hull adherence and even germination) are extremely important aspects for barley improvement. While considerable progress has been achieved, much remains to be done in terms of improving the quality and production of malting barley. Quality of barley significantly affects its end utilization. Statistical data indicate that approximately 19% of total barley produced is used in malting process, 8% is consumed as food, 2% in industrial processes and about 73% is used for animal feed due to inadequate malting quality (e-malt.com, 2007). The issues linked with germination of malting barley have acquired substantial global attention for the last few years. It is evident from the literature that storage conditions and pre-harvest sprouting have major consequences on germination. The underlying causes for varietal differences in these characteristics is still unclear. Secondary dormancy greatly reduces the germination and marketability of grains used for malting purposes. Therefore, there is dire need to address this issue that malting barley sustains its germination without prolonged dormancy and pre-harvest sprouting.

2. Seed dormancy

2.1. Seed dormancy: Definition

Seed dormancy is a common characteristic of wild plants which ensures their continued existence/survival under unfavourable conditions, decreases competition with other species and prevents damage to seedlings from out-of-season germination of the seed. Domesticated species, on the other hand, are selected for uniform germination and rapid seedling establishment often leading to selection of genotypes with less dormancy. This can lead to pre-harvest sprouting (PHS), a phenomenon in which the seed germinates on the parent plant causing extensive loss of grain quality to crops like wheat, barley and maize (Bewley & Black, 1994).

Seed dormancy is defined as *"inhibition of germination of an intact viable seed under favourable conditions"* (Hilhorst, 1995; Li & Foley, 1997; Bewley, 1997). The germination block has developed in a different way from one species to another depending upon their habitat and conditions of growth. These dormancy mechanisms have evolved because these germination blocks have been operated in a variety of climates and habits. In light of these complex nature of germination blocks, another definition of dormancy has been defined as, a *"dormant seed cannot germinate in a specified period of time under any combination of conditions that are otherwise sufficient for its germination"* (Baskin & Baskin, 2004). It is reported that dormancy must not be linked with lack of germination, but dormancy is the combination of characteristics of the seed which decide physical and environmental circumstances needed for germination (Finch-Savage & Leubner-Metzger, 2006). Germination can be defined as appearance of radicle from seed coat. The requirement of germination may include one or more of the processes like mobilization of stored food, overcoming the physical barrier by activation of cell wall degrading enzymes followed by resumption of active growth by cell elongation and division (Finkelstein et al., 2008).

2.2. Classification of seed dormancy

Although almost all kinds of dormancy cause delay in germination, the principal of this delay may vary from species to species. The variation can be due to embryonic immaturity or due to the existence of physical or physiological constraints caused by the presence of a hard seed coat or some inhibitory chemicals that interfere with embryo growth (Finch-Savage & Leubner-Metzger, 2006). Dormancy can be primary dormancy that is acquired in the later developmental phases of embryo development and seed maturity. There are also conditions in which after-ripened, imbibed seeds enter into secondary dormancy when exposed to unfavourable temperature, light or low moisture conditions (Bewley, 1997).

Despite the progress in understanding the mechanisms controlling dormancy, it can be treated as the least recognized event (Finch-Savage & Leubner-Metzger, 2006). Both physiologists and ecologists have studied the factors controlling dormancy but the outcome is far from clear due to the fact that dormancy is affected by numerous environmental conditions (an ecologist's dilemma) and the model species like Arabidopsis studied by molecular physiologists and geneticists tend to have a very shallow dormancy (Walck et al., 2005). The molecular controls that regulate dormancy can be of two different components i.e., an embryo or a seed coat. However, dormancy is a entire seed trait and on this basis, can be classified into the five classes namely physiological, morphological, morpho-physiological, physical and combinatorial dormancy (Nikolaeva, 1969; Baskin & Baskin, 2004; Finch-Savage & Leubner-Metzger, 2006).

2.3. Factors affecting dormancy

Dormancy is affected by various factors and the potential regulators are identified by their effect on depth of dormancy or by analysis of genetic lines that have varying levels of dormancy. The factors that affect dormancy are classified into two broad categories, embryo- and seed coat imposed dormancy. A hard seed coat manifests its effect on dormancy by prevention of water uptake during imbibition (waxy or lignified tissues in legume seeds), mechanical constraint due to hard seed coat (nuts) or endosperm (lettuce) causing inhibition of radicle protrusion, interference with gas exchange (cocklebur) and retention of inhibitors (*Xanthium*) and production of inhibitors like abscisic acid (ABA). Genetic variation in seed coat components such as testa layer, pericarp and pigmentation also cause altered dormancy and seed longevity (Debeaujon et al., 2000; Groos et al., 2002; Sweeney et al., 2006). Pigmented seeds are generally more dormant although hormone levels and their sensitivity to them may increase dormancy of non-pigmented seeds (Gale et al., 2002; Walker-Simmons, 1987; Flintham, 2000). Many nitrogenous compounds like nitrite (NO_2^-), nitric oxide (NO), and nitrate (NO_3^-) cause dormancy release. NO could promote germination by cell wall weakening and instigating vacuolation (Bethke et al., 2007). Genomic studies in rice to identify loci controlling seed colour, dormancy and shattering resistance show a tight linkage between the responsible genes and single locus can also control these traits (Ji et al., 2006).

Embryo dormancy is controlled by inherent characteristics of the embryo. The presence or absence of embryo dormancy has mainly been attributed to the content and sensitivity of phytohormones ABA and gibberellic acid (GA) (Bewley, 1997). Dormancy and germination are also affected by environmental factors such as light, moisture and temperature (Borthwick et al., 1952; Gutterman et al., 1996). The intensity of dormancy in the mature seed and its onset during seed development vary considerably due to genotype by environmental interaction during the entire process of seed development (Corbineau et al., 2000; Crome et al., 1984; Bewley, 1997;).

2.4. Hormonal control of dormancy

The plant hormone abscisic acid is required for setting dormancy during embryo maturation and its deposition associate with the commencement of primary dormancy (Kermode, 2005). Another plant hormone, gibberellic acid is antagonistic in action to ABA. Gibberellins promote post-germinative growth by activating hydrolyzing enzymes that break cell walls, mobilize seed storage reserves and stimulate embryo cell expansion (Bewley, 1997). Ethylene also promotes germination by antagonizing ABA signalling. Ethylene receptor mutants have higher ABA content and are hypersensitive to ABA (Ghassemian et al., 2000; Beaudoin et al., 2000; Chiwocha et al., 2005). Plant steroidal hormones, brassinosteroids, enhance the germination potential of embryos in a GA-independent manner (Leubner-Metzger, 2001). The germination completion and establishment of seedling is accomplished by Auxin (Carrera et al., 2007; Ogawa et al., 2003; Liu et al., 2007a). Auxin accumulates at the radicle tip during embryo development and in seeds after imbibition (Liu et al., 2007a). Although various hormones may affect dormancy and germination, the general consensus is that ABA is the primary mediator of dormancy (Koornneef et al., 2002; Holdsworth et al., 2008; Finkelstein et al., 2008).

2.5. ABA and GA regulate dormancy and germination

The functions of ABA in dormancy maintenance and initiation are firmly established and widely reviewed (Koornneef et al., 2002; Finch-Savage & Leubner-Metzger, 2006; Finkelstein et al., 2008). In cereals like wheat, barley and sorghum, ABA controls the onset of dormancy

(Walker-Simmons, 1987; Jacobsen et al., 2002). Genetic studies show that the *de novo* synthesis of ABA in embryo or endosperm is required to induce dormancy (Nambara & Marion-Poll, 2003). Other studies with ABA-deficient mutants have suggested that ABA in the embryos and not the maternal ABA is crucial for induction of dormancy (Karssen et al., 1983). Dormancy may be maintained by renewed post-imbibition synthesis of ABA (LePage-Degivry & Garello, 1992; Ali-Rachedi et al., 2004). The reduction in seed dormancy has been seen for ABA biosynthetic enzymes, that have ABA sequestration with expressed antibodies in the seeds and in seeds that are treated with chemicals for inhibition of ABA biosynthesis (Nambara & Marion-Poll, 2003; Lin et al., 2007). The content of ABA and resulting dormancy are controlled by interaction of ABA biosynthetic and ABA catalyzing enzymes. The most critical enzyme in ABA biosynthesis is the 9-*cis*-epoxycarotenoid dioxygenase (NCED) that is essential for ABA synthesis in endosperm and embryo (Lefebvre et al., 2006). Rate-limiting enzyme during ABA biosynthesis, NCED regulates ABA biosynthesis during induction of secondary dormancy (Leymarie et al., 2008). During the transition from embryo maturation to germination, ABA is catabolised by ABA 8′-hydroxylases which are encoded by cytochrome P450 CYP707A gene family causing a decline in dormancy (Okamoto et al., 2006). Imbibition of embryos in water also causes leaching of ABA resulting in reduced dormancy (Suzuki et al., 2000). After-ripening, which is occurring during dry storage of seeds, causes a decline in embryo ABA content and sensitivity (Grappin et al., 2000). In a study conducted on pre-harvest sprouting (PHS) in susceptible and resistant wheat cultivars, after-ripening occured prior to harvest ripeness in the majority of PHS-susceptible cultivars, whereas it was slowest in cultivars that were most PHS-resistant. However, no direct relationship could be found between timing of caryopsis after-ripening and dormancy or ABA responsiveness in wheat (Gerjets et al., 2009).

ABA content as well as ABA sensitivity are critical components of embryo dormancy. ABA-insensitive mutants that are deficient in ABA perception or signalling have lower dormancy and exhibit viviparous germination (Koornneef et al., 1984; Robichaud & Sussex, 1986; Koornneef et al., 1989). Analysis of sprouting-susceptible and sprouting-resistant cultivars of wheat for ABA content and ABA sensitivity showed larger differences in ABA sensitivity than ABA content measured by capability of ABA to block embryo germination (Walker-Simmons, 1987).

The role of GA in modulating dormancy is highly debated (Finkelstein et al., 2008). The treatment with GA may not direct germination in few species or in fully dormant seeds of Arabidopsis. The decline of ABA content is usually needed prior to embryo GA content or sensitivity to the hormone increases (Ali-Rachedi et al., 2004; Jacobsen et al., 2002). After-ripening, which leads to a decline in ABA content and ABA sensitivity, results in increased sensitivity to GA and light in Arabidopsis (Derkx & Karssen, 1993). So the ratio of ABA to GA seems to be critical, where a higher content of ABA overrides the growth-promoting effect of GA. In cereals, although the GA signalling components seem to be similar to dicots, redundant GA signalling pathways may exist. This is evident from the fact that in rice, the mutation in the only known receptor of GA, *Gibberellin-Insensitive Dwarf 1* (*GID1*) leads to decreased α-amylase production (Ueguchi-Tanaka et al., 2005); however mutating all three homologues of *GID1* in Arabidopsis inhibits germination (Willige et al., 2007). Therefore, it can be concluded that the embryo dormancy in case of cereals, for the most part, is controlled by ABA content and sensitivity.

2.6. Effect of light on dormancy occurs through ABA and GA metabolism

The role of light in regulation of dormancy was first identified when germination was induced by exposing the dark-imbibed seeds with red (R) light pulse and the successive far-red (FR) light pulse cancelled the effect of red light (Borthwick et al., 1952). This response is mediated by the R/FR phytochromes, UV-A/blue light receptor cryptochromes, the phototropins and the recently identified blue light receptor zeitlupes (Bae & Choi, 2008).

The induction of germination by red light can be substituted by the application of GA (Kahn et al., 1957), whereas red light application do not induce germination in mutants deficient in GA (Oh et al., 2006). Toyomasu et al., (1998) reported that the GA biosynthetic gene's expression encoding GA3ox (*LsGA3ox1* in lettuce and *AtGA3ox1* and *AtGA3ox2* in Arabidopsis) is generated by R light and its activation is inhibited by FR light. On the other hand, transcripts of a GA-deactivating gene *GA2ox* (*LsGA2ox2* in lettuce and *AtGA2ox2* in Arabidopsis) are reduced by R light (Yamauchi et al., 2007; Oh et al., 2006; Nakaminami et al., 2003; Seo et al., 2006).

Similar to modulation of GA content, ABA biosynthetic and deactivating enzymes are also regulated by light. Genes encoding ABA biosynthetic enzymes NCED (*LsNCED2* and *LsNCED4* in lettuce and the Arabidopsis *AtNCED6* and *AtNCED9*) and zeaxanthin epoxidase (AtZEP/AtABA1 in Arabidopsis) are reduced by R light treatment (Seo et al., 2006; Sawada et al., 2008; Oh et al., 2007) whereas, transcript levels of ABA-deactivating genes encoding CYP707A (*LsABA8ox4* in lettuce and CYP707A2 in Arabidopsis) are elevated by R light (Sawada et al., 2008; Oh et al., 2007; Seo et al., 2006).

The phytochromes regulate the levels of ABA and GA by one of the interrelating proteins *PHYTOCHROME INTERACTING FACTOR3-LIKE 5* (*PIL5*) which belongs to a family of helix-loop-helix (bHLH) family of proteins containing 15 members (Yamashino et al., 2003; Toledo-Ortiz et al., 2003;). Studies of *PIL5* over-expressing and mutant lines show that it regulates ABA and GA content by regulating their metabolic genes (Oh et al., 2006).

3. Molecular networks regulating dormancy

3.1. Perception and transduction of ABA signal

3.1.1. ABA receptors

Physiological studies in different plant species indicate that accumulation of ABA is required for induction and maintenance of dormancy (Finkelstein et al., 2008). The perception of ABA and its downstream signalling to inigtiate ABA-regulated responses is an area of active research. Various lines of evidence suggest multiple sites of ABA perception, thus, multiple ABA receptors (Allan & Trewavas, 1994; Gilroy & Jones, 1994; Huang et al., 2007). The first

ABA-specific binding protein, a 42 kDa ABAR, was identified and isolated from *Vicia faba* leaves and the pretreatment of their guard cell protoplasts with a monoclonal antibody against the 42 kDa protein reduced ABA induced phospholipase D activity in a manner that was dose-dependent (Zhang et al., 2002). Another 52kDa protein, ABAP1 was shown to bind ABA and was up-regulated by ABA in barley aleurone layer tissue (Razem et al., 2004). The ABA "receptor", Flowering Time Control Locus A (FCA) in Arabidopsis was identified based on its high sequence similarity to barley ABAP1 and was shown to bind ABA and affect flowering (Razem et al., 2006). Another ABA receptor from Arabidopsis, the Magnesium Protoporphyr-in-IX Chelatase H subunit (CHLH) regulates classical ABA-regulated processes like stomatal movements, post germination growth and seed germination (Shen et al., 2006).The CHLH also shared very high sequence similarity to ABAR (Shen et al., 2006). In 2008, questions about FCA being a receptor for ABA arose in both the laboratory of the original authors and, independently, in laboratories in New Zealand and Japan. This culminated in the simultaneous publication of a letter questioning the original results (Risk *et al.* 2008) and a retraction of the claim that FCA was an ABA receptor (Razem et al., 2006). Subsequent studies have confirmed that the findings of Razem et al., (2006) were not reproducible (Risk et al., 2009; Jang et al., 2008). Questions have also been raised regarding CHLH and its effect on feedback regulation of ABA synthesis and the apparent lack of a mechanism for its ABA receptor function (Shen et al., 2006; Verslues & Zhu, 2007). CHLH binding to ABA was proven using more than one method (Wu et al., 2009). Yet the barley homologue of CHLC (magnesium chelatase 150 kD subunit) does not bind ABA (Muller & Hansson, 2009). Two classes of plasmamembrane ABA receptor, a G-protein-coupled receptor (GPCR), the GCR2, and a novel class of GPCR, the GTG1 and the GTG2 have been discovered. They regulate major ABA responses such as seed germination, seedling growth and stomatal movement (Liu et al., 2007b; Pandey et al., 2009). However, the GCR2 mediation of ABA-controlled seed germination and post-germination growth are controversial as the ABA-related phenotypes are lacking or weak in *gcr2* mutants (Gao et al., 2007; Guo et al., 2008). GTGs regulate ABA signalling positively and interact with the only Arabidopsis G-protein α-subunit, GPA1, which can negatively regulate ABA signalling by nullyfying the activity of GTG-ABA binding (Pandey et al., 2009). The ABA insensitive mutants *abi1* and *abi2* belong to Mg^{2+}- and Mn^{2+}-dependent serine-threonine phosphatases type 2C (PP2Cs) and are known to be negative regulators of ABA signalling (Merlot et al., 2001; Gosti et al., 1999; Rodriguez et al., 1998; Meyer et al., 1994). The 14 member gene family of Regulatory Components of ABA Receptor (RCARs), which interact with ABI1 and ABI2, bind ABA, mediate ABA-dependent inactivation of ABI1 and ABI2 *in vitro* and antagonize PP2C action *in planta (Ma et al., 2009)*. PYRABACTIN RESISTANCE 1 (PYR/PYL family of START proteins) were shown to inhibit the PP2C mediated ABA signaling (Park, 2009). In Arabidopsis, the PYR/PYL/RCAR family proteins constitute the major in vivo phosphatase 2C-interacting proteins (Noriyuki et al., 2010). The crystal structure of Arabidopsis PYR1 indicated that the molecule existed as a dimer, and the mechanism of its binding to ABA in one of the PYR1 subunits was recently established (Nishimura et al., 2009; Santiago et al., 2009). Finally, the whole ABA signalling cascade that includes PYR1, PP2C, the serine/threonine protein kinase SnRK2.6/OST1 and the transcription factor ABF2/AREB1 was reconstituted *in vitro* in plant protoplasts resulting in ABA responsive gene expression (Fujii et al., 2009).

3.1.2. ABA signalling components

To identify the different ABA signalling components, various Arabidopsis mutants were screened for insensitivity to ABA for germination and were termed ABA insensitive (abi) (Koornneef et al., 1984; Finkelstein, 1994). The ABI1 and ABI2 encoded protein phosphatase 2C (type 2C phosphatases, PP2C) regulate ABA signalling (Leung et al., 1997). ABI3, ABI4 and ABI5 control mainly seed related ABA responses (Parcy et al., 1994; Finkelstein & Lynch, 2000).

The process of dormancy initiates during early seed maturation and continues until the seed matures completely (Raz et al., 2001). In Arabidopsis, the seed maturation and induction of dormancy is mainly controlled by four transcription factors namely FUSCA3 (FUS3), ABSCISIC ACID INSENSITIVE 3 (ABI3), LEAFY COTYLEDON 1 (LEC 1) and LEC 2 (Stone et al., 2001; Baumlein et al., 1994; Giraudat et al., 1992; Lotan et al., 1998). The plant specific transcription factors with the conserved B3-binding domain include ABI3, FUS3 and LEC2 (Stone et al., 2001). LEC1 encodes the HAP3 subunit of a CCAAT-binding transcription factor CBF (Lotan et al., 1998). Common mutant phenotypes such as decreased dormancy at maturation occur due to abi3, lec1, lec2 and fus3 and they affect seed maturation severely (Raz et al., 2001) as well as cause reduced expression of seed storage proteins (Gutierrez et al., 2007). A study, using Arabidopsis cultivars that differed in dormancy, showed no correlation between LEC1, FUS3, ABI3 and Em expression and dormancy (Baumbusch et al., 2004). Although all four genes affect embryo maturation, they also play a unique role in regulating each other's functionality and expression pattern (Holdsworth et al., 2008). FUS3 controls formation of epidermal cell identity and embryo derived dormancy (Tiedemann et al., 2008). Loss of LEC1 causes germination of excised embryos similar to lec2 and fus3 mutants (Raz et al., 2001). LEC2 controls the transcription program during seed maturation and affects DELAY OF GERMINATION 1 (DOG1), the first seed dormancy protein accounting for variation in natural environment as identified by quantitative trait loci (QTL) analysis (Bentsink et al., 2006; Braybrook et al., 2006). Both LEC1 and LEC2 regulate the expression of FUS3 and ABI3 (Kroj et al., 2003; Kagaya et al., 2005). ABI3 and FUS3 positively auto-regulate themselves and each other creating a feedback loop (To et al., 2006). Interestingly, none of these four transcription factors (LEC1, FUS3, ABI3 and LEC2) contains motifs to interact with an ABA response element (ABRE), but do contain a B3 domain that interacts with the RY motif present in the promoters of genes that produce RNA during the late maturation phase of the seed (Ezcurra et al., 1999; Reidt et al., 2000; Monke et al., 2004; Braybrook et al., 2006). The transcription factor ABSCISIC ACID INSENSITIVE 5 (ABI5) is a basic leucine zipper (bZIP) domain containing protein that interacts with ABRE and activats ABA-mediated transcription in seeds (Finkelstein & Lynch, 2000; Carles et al., 2002). ABI3 activates RY elements, physically interacts with ABI5 and this physical interaction seems to be necessary for ABA-dependent gene expression (Nakamura et al., 2001).

Although much information on dormancy regulation is available for dicots like Arabidopsis, the molecular control of dormancy in cereals is not very clear. One of the key genes in regulating seed maturation, dormancy and desiccation in maize is Viviparous1 (VP1), an ortholog to ABI3 in Arabidopsis (McCarty et al., 1989; McCarty et al., 1991; Giraudat et al., 1992). It is also responsible for transcriptional control of the LATE EMBRYOGENESIS

ABUNDANT (LEA) class of proteins (Nambara et al., 1995; Nambara et al., 2000). VP1 is involved in root growth-related interaction between ABA and auxin (Suzuki et al., 2001). QTL analysis showed VP1 to be responsible for seed dormancy and PHS (Flintham et al., 2002; Lohwasser et al., 2005). VP1 is responsible for controlling embryo maturation and dormancy as well as inhibition of germination (McCarty & Carson, 1991; Hoecker et al., 1995). Like ABI3, ABI5 and VP1 interac to regulate embryonic gene expression and sensitivity of seed to ABA (Lopez-Molina et al., 2002). VP1/ABI3 has been cloned from various dicot and monocot species (Hattori et al., 1994; Jones et al., 1997; Rohde et al., 2002) and contains three basic domains designated B1, B2 and B3 and a N-terminal acidic domain (A1) (Giraudat et al., 1992). The A1 domain is responsible for ABA-mediated transcriptional activation, B2 for ABRE-mediated transcriptional activation and B3 for RY/G-box interaction (Hoecker et al., 1995; Ezcurra et al., 1999). VP1/ABI3 is also alternatively spliced in various plant species and its mis-splicing causes PHS in wheat (McKibbin et al., 2002; Wilkinson et al., 2005; Gagete et al., 2009). ABI5 undergoes alternative splicing forming two variants which interact with each other and each having distinct binding affinity to VP1/ABI3 (Zou et al., 2007). In barley, ABA-dependent up-regulation of ABI5 is positively regulated by a feed-forward mechanism that involves ABI5 itself and VP1 (Casaretto & Ho, 2005).

Our work on FCA and FY, two key components in regulation of flowering, suggest that commonalities exist in germination and flowering pathways. The transcript levels of barley FCA are positively correlated to dormant state of the embryos and are involved in regulation of VP1 and Em gene promoters (Kumar et al., 2011). The Arabidopsis FY, which regulates the autonomous floral transition pathway through its interaction with FCA, is also involved in seed germination in Arabidopsis (Jiang et al., 2012). The fy-1 mutant has lower ABA sensitivity and may be involved in development of dormancy (Jiang et al., 2012). These reports suggest a very prominent role of transcriptional regulation in fine tuning ABA responses.

3.2. Inhibition of GA signalling by DELLA proteins

Components of GA signalling regulate seed germination (Peng & Harberd, 2002). Nuclear transcriptional regulators, the DELLA proteins, control GA signalling (Itoh et al., 2002; Richards et al., 2000; Wen & Chang, 2002; Dill et al., 2001). DELLA proteins are negative regulators of GA signalling (Wen & Chang, 2002). Arabidopsis has five DELLA proteins (GA-INSENSITIVE [GAI], REPRESSOR OF GA1-3 [RGA], RGA-LIKE1 [RGL1], RGL2, and RGL3), while rice SLENDER1 (SLR1) and other species such as barley SLENDER1 (SLN1), maize, and wheat have only one DELLA protein (Dill et al., 2001; Chandler et al., 2002; Itoh et al., 2002; Peng & Harberd, 2002). Downstream of the DELLA proteins, GA regulates Myb-like (GAmyb) transcription factor binding to promoter of α-amylase genes (Gubler et al., 1995). The GA-signal is recepted by a soluble GA receptor which has homology to GA-INSENSITIVE DWARF1 (GID1), a human hormone-sensitive lipase (Ueguchi-Tanaka et al., 2005). The bioactive GAs bind to GID1 which in turn promotes interaction between GID1 and the DELLA-domain of DELLA protein (Willige et al., 2007; Ueguchi-Tanaka et al., 2007). This interaction enhances the affinity between DELLA-GID1-GA complex and a specific SCF E3 ubiquitin–ligase complex, SCFSLY1/GID2 which involves the F-box proteins AtSLY1 and OsGID2 in

Arabidopsis and rice, respectively (Sasaki et al., 2003; McGinnis et al., 2003; Willige et al., 2007; Griffiths et al., 2006). The ubiquitinylation and subsequent destruction of DELLAs is promoted by SCFSLY1/ GID2 through the 26S proteasome (Fu et al., 2002; McGinnis et al., 2003; Sasaki et al., 2003). The DELLA genes are transcriptionally controlled by the light-labile transcription factor PIL5 which increases the transcription of GAI and RGA genes by binding to its promoters on the G-Box (Oh et al., 2007).

DELLA degradation is GA-dependent and is inhibited by ABA in barley and by both ABA and salt (NaCl) in Arabidopsis (Gubler et al., 2002; Achard et al., 2006). Plant development through the two independent salt-activated hormone signalling pathways (ABA and ethylene) integrates at the level of DELLA function (Achard et al., 2006). DELLA also affects flowering in an ABA-dependent manner (Achard et al., 2006); however, its function in regulation of dormancy and germination is not clear. Germination in tomato, soybean and Arabidopsis is not dependent on down-regulation of DELLA genes (Bassel et al., 2004). Despite a high content of RGL2, the DELLA protein that specifically represses seed germination, Arabidopsis sly1 mutant seeds can germinate (Ariizumi & Steber, 2007). Far-red light is known to inhibit germination through DELLA dependent induction of ABI3 activity and ABA biosynthesis while DELLA mediates expansion of cotyledon leading to breaking the coat-imposed dormancy (Penfield et al., 2006; Piskurewicz et al., 2009).

4. Epigenetic regulation of dormancy related genes

Despite the lack of complete information about ABA signalling, it is amply clear that ABA responses are regulated by transcriptional regulation, except for the quick responses in stomatal closure (Wasilewska et al., 2008). Besides transcriptional regulation, ABA mediates epigenetic regulation to control plant responses (Chinnusamy et al., 2008). ABA-mediated epigenetic regulation of gene expression in seeds is now being studied extensively. Polycomb group-based gene imprinting and DNA methylation/demethylation control seed development in plants (Eckardt, 2006). Seed specific physiological processes like dormancy and germination are being studied in the context of epigenetic regulation. A cDNA-AFLP-based study showed epigenetic regulation of transcripts during barley seed dormancy and germination (Leymarie et al., 2007). During seed development and germination inhibition, gene regulation is also regulated by ABA through transcription factors such as ABI3, VP1, LEC2, FUS3 as well as the APETELA2 (ABI4), HAP3 subunit of CCAAT binding factor (LEC1) and the bZIP (ABI5) (Finkelstein et al., 2002). ABA regulates the B3 domain transcription factors through PICKLE (PKL) which encodes putative CHD3 type SWI/SNF-class chromatin-remodeling factor (Ogas et al., 1999). ABA-mediated stress responses occur through Histone Deacetylase (HDACs)-dependent chromatin modifications and ATP-dependent chromatin remodelling complexes that include SWI3-like proteins (Wu et al., 2003; Rios et al., 2007). Stress-related memory is also inherited through epigenetic mechanisms (Boyko et al., 2007). ABA also regulates non-coding small RNAs (siRNA and miRNA) that can regulate DNA methylation resulting in epigenetic changes (Bond & Finnegan, 2007; Yang et al., 2008).

5. Tillering and bud dormancy

Tillering is a key agronomic trait contributing to grain yield. Tillers are formed from axillary buds that grow independent of the main stem. The levels of dormancy in buds determine the timing and extent of tillers in most monocot crops. Various proteins such as MONOCULM 1 (MOC1) (Li et al., 2003) have been implicated in regulation of bud dormancy but recent studies suggest the involvement of autonomous pathway (flowering) genes in regulation of bud dormancy. The first clue regarding the commonality between factors controlling flowering and bud dormancy arose from environmental signals that regulated them (Chouard, 1960). The signalling events responsible for regulation of flowering and bud dormancy converge on FLOWERING LOCUS T (FT) (Bohlenius et al., 2006). Day length is an important determinant in regulation of flowering acting through its photoreceptor PHYTOCHROME A (PHYA). PHYA affects the floral induction pathway through its effect on CONSTANS (CO), a gene involved in flowering pathway, which in turn affects FT (Yanovsky & Kay, 2002). FT is negatively regulated by FLC which regulates temperature-dependent seed germination in Arabidopsis (Helliwell et al., 2006; Chiang et al., 2009). FCA and FVE regulate FT under high and low temperatures in a FLC-dependent manner (Sheldon et al., 2000; Blazquez et al., 2003). The transcript levels of *FCA* have also been correlated to bud dormancy in poplar (Ruttink et al., 2007). Although limited, the information regarding the intricate network of signalling events that regulate the two most important events, namely the transition from vegetative to reproductive state, and from non-germinated to germinated state suggests some common factors (Horvath, 2009).

6. Breeding for pre-harvest resistance in barely

Seed dormancy is a quantitatively inherited trait in several plant species such as rice, popular, Arabidopsis, wheat and barley (Ullrich et al., 1996; Li et al., 2004). In barley, seed dormancy and germination have been important breeding objectives since its domestication and malt utilization. Malting barley must rapidly germinate upon imbibition. Endosperm starch and proteins hydrolysis within 3 to 4 days is an important characteristic for malting quality in barley. To assure rapid and complete germination for malting industry, barley breeders have stringently selected against seed dormancy resulting in barley varieties that are highly susceptible to pre-harvest spouting after early fall rains or heavy dew, which is an undesirable trait (Prada et al., 2004). A moderate level of seed dormancy is desirable for proper malting. In order to achieve suitable level of seed dormancy, several studies reported seed dormancy QTLs in barley (Edney & Mather, 2004; Zhang et al., 2005), different dormancy genes however responsible in different population of various pedigrees. Levels of seed dormancy that vary in different genetic backgrounds are also affected by environmental factors and their interaction with genetic factors. Various studies have identified the major QTLs (SD1 and SD2) that can be used in combination with other minor QTL of local germplasm to achieve moderate level of seed dormancy for malting barley (Li et al., 2004). Few QTL identified in barley for dormancy and preharvest sprouting are listed in Table 1. In addition hormonal cross talk can be explored for seed dormancy and germination as breeding prospect for better barley values and end utilization.

Chromosome	Marker interval	Variability (%)	References
Cross: Setptoe x Morex			
5H	*Ale* - ABC324	50	Ullrich et al., 1993
5H	MWG851D - MWG851B	15	Obethur et al., 1995
7H	*Amy2 - Ubi1*	5	Han et al., 1996 Ullrich et al., 2002
4H	WG622 - BCD402B	5	Gao et al., 2003
Cross: Chebec x Harrington			
5H	CDO506 - GMS1	70	Li et al., 2003
Cross: Hordeum spontaneum (Wadi Qilt) x *Hordeum vulgare* (Mona)			
$1H_1$	ABC160-3	13	
$5H_2$	BMAG812-1 – E35M59mg-4	14	
$1H_2$	EMBAC659-3 – EE38M55ob-1	45	
$7H_1$	AF22725-3 – BMAG341A-2	13	Zhang et al., 2005
$7H_2$	BMAG135-4 – HVPR1B-2	39	
$1H_1$	EMBAC659-3 – EE38M55ob-1	50	
Cross: Stirling x Harrington			
1Hq	Hvglvend – Awbms80	1.6	
2Hqa	GBMS244 – Emag174	-	
3Hqa	GBM1043 – Bmag0013	2.2	Li et al., 2003,
4Hqa	GBM1501 – Bmag741	-	Bonnardeaux et al., 2008
5Hqa	Bmag0337 – GBM1399	3.7	
5Hqb	Scsst09041a – scssr03901	52.2	
Cross: Harrington x TR306			
1HL	iPgd2 – TubA2	10	
2HS	ABC019 – ABG716	7	
2HC	MWG865	8	
3HL	ABG609B – MWG838	13	Ullrich et al., 2009
5HL	MWG602 – ABC718	40	
7HS	dRPG1 – ABG077	6	
7HC	MWG003 – Ris15	7	
Cross: Triumph x Morex			
1HS	GMS21	10	Ullrich et al., 2009
3HL	E39M49_j – E39M48_c	13	Prada et al., 2004

Chromosome	Marker interval	Variability (%)	References
5HC	E39M49_f – MWG522	54	
7HC	E32M48_c – E39M48_p	7	
7HL	E37M60_g	7	
Cross: BCD47 x Baronesse			
1H	Bmag504 - Bmag032	10	
4H	HvSnf2 – HvAmyB	9	
5H	Bmag222 – GMS001	34.5	Castro et al., 2010
6H	Bmag500 - Bmag009	9	
7H	Bmag120 – Ris44	23	
Cross: ND24260 x Flagship			
3H	bPb – 0619	6	
3H	bPb – 2630	4	
4H	bPb – 9251	4	
5H-2	bPb – 9191	15	
5H-2	bPb – 5053	31	Hickey et al., 2012
5H-2	bPb – 1217	35	
5H-2	bPb – 1217	28	
6H-2	bPb - 1347	4	

Table 1. Dormancy and preharvest sprouting related QTLs in barley.

7. Future perspective

The plethora of information on molecular control of dormancy and germination is ever increasing with studies performed on model plants. Little information is available from agriculturally important crops such as wheat and barley as they are tedious systems due to their genome complexity and ploidy levels. However, these economically important crops do bring out the unique variations of the biological systems that improve our understanding.

The recent pieces of evidence from our studies in barley and Arabidopsis (Kumar et al., 2011; Jiang et al., 2012) lay a foundation for looking deeply into the bigger picture involving flowering and dormancy as connected pathways. Genetic studies in Arabidopsis also identified DOG1, a key component in dormancy pathway, as quantitative trait loci for flowering (Atwell et al., 2010). The improvements in next generation sequencing and its decreasing cost has made it the technology of choice for looking at entire genomes for various transcriptome and epigenetic studies in crop plants. A refocused approach using all interconnected pathways and improved technologies to study them will certainly enhance our understanding of dormancy and germination as well as flowering and in turn promote crop improvement.

Abbreviations

ABA Abscisic Acid

ABAP1 ABA Binding Protein 1

ABI3 Abscisic Acid Insensitive 3

ABI5 Abscisic Acid Insensitive 5

DOG1 DELAY OF GERMINATION 1

FCA Flowering Time Control Protein A

FLC Flowering Control Locus C

FT Flowering Locus T

GA Gibberellic Acid

GID1 GA-INSENSITIVE DWARF1

LEA Late Embryogenesis Abundant

LEC 1 LEAFY COTYLEDON 1

PHS Pre-harvest Sprouting

SLN1 Slender 1 (DELLA protein)

VP1 Viviparous 1

EM Early Methionine

Acknowledgements

The authors are grateful to Dr. Robert Hill and Dr. Derek Bewley for their expert opinion and advice for preparation of this manuscript. This book chapter has been taken from Dr. Santosh Kumar's PhD thesis entitled "Molecular and Physiological Characterization of the Flowering Time Control Protein, HvFCA and its Role in ABA Signalling and Seed Germination" submitted to the faculty of graduate studies, University of Manitoba.

Author details

Santosh Kumar[1], Arvind H. Hirani[1], Muhammad Asif[2*] and Aakash Goyal[3]

*Address all correspondence to: asifrana@gmail.com

1 Department of Plant Science, University of Manitoba, Winnipeg, Manitoba, Canada

2 Agricultural, Food and Nutritional Science, Agriculture/Forestry Centre, Univ. of Alberta, Edmonton, AB, Canada

3 Bayer Crop Science, Saskatoon, Canada

References

[1] Achard, P, Cheng, H, De Grauwe, L, Decat, J, Schoutteten, H, Moritz, T, Van Der Straeten, D, Peng, J. R, & Harberd, N. P. (2006). Integration of plant responses to environmentally activated phytohormonal signals. *Science*, 311, 91-94.

[2] Ali-rachedi, S, Bouinot, D, Wagner, M. H, Bonnet, M, Sotta, B, Grappin, P, & Jullien, M. (2004). Changes in endogenous abscisic acid levels during dormancy release and maintenance of mature seeds: studies with the *Cape Verde Islands* ecotype, the dormant model of *Arabidopsis thaliana*. *Planta*, 219, 479-488.

[3] Allan, A. C, & Trewavas, A. J. (1994). Abscisic-acid and gibberellin perception- Inside or out. *Plant Physiology*, 104, 1107-1108.

[4] Ariizumi, T, & Steber, C. M. (2007). Seed germination of GA-insensitive sleepy1 mutants does not require RGL2 protein disappearance in Arabidopsis. *The Plant Cell*, 19, 791-804.

[5] Atwell, S, Huang, Y. S, Vilhjalmsson, B. J, Willems, G, Horton, M, Li, Y, Meng, D, Platt, A, Tarone, A. M, Hu, T. T, Jiang, R, Muliyati, N. W, Zhang, X, Amer, M. A, Baxter, I, Brachi, B, Chory, J, Dean, C, Debieu, M, De Meaux, J, Ecker, J. R, Faure, N, Kniskern, J. M, Jones, J. D, Michael, T, Nemri, A, Roux, F, Salt, D. E, Tang, C, Todesco, M, Traw, M. B, Weigel, D, Marjoram, P, Borevitz, J. O, Bergelson, J, & Nordborg, M. (2010). Genome-wide association study of 107 phenotypes in *Arabidopsis thaliana* inbred lines. *Nature*, 465, 627-631.

[6] Badr, A, Muller, K, Schafer-pregl, R, Rabey, H. E, Effgen, S, Ibrahim, H. H, Pozzi, C, Rohde, W, & Salamini, F. (2000). On the origin and domestication history of barley (*Hordeum vulgare*). *Molecular Biology and Evolution*, 17, 499-510.

[7] Bae, G, & Choi, G. (2008). Decoding of light signals by plant phytochromes and their interacting proteins. *Annual Review of Plant Biology*, 59, 281-311.

[8] Baskin, J. M, & Baskin, C. C. (2004). A classification system for seed dormancy. *Seed Science Research*, 14, 1-16.

[9] Bassel, G. W, Zielinska, E, Mullen, R. T, & Bewley, J. D. (2004). Down-regulation of *DELLA* genes is not essential for germination of tomato, soybean, and Arabidopsis seeds. *Plant Physiology*, 136, 2782-2789.

[10] Baumbusch, L. O, Hughes, D. W, Galau, G. A, & Jakobsen, K. S. (2004). *LEC1, FUS3, ABI3* and *Em* expression reveals no correlation with dormancy in Arabidopsis. *Journal of Experimental Botany*, 55, 77-87.

[11] Baumlein, H, Misera, S, Luerssen, H, Kolle, K, Horstmann, C, Wobus, U, & Muller, A. J. (1994). The *Fus3* gene of *Arabidopsis Thaliana* Is a regulator of gene-expression during late embryogenesis. *The Plant Journal*, 6, 379-387.

[12] Beaudoin, N, Serizet, C, Gosti, F, & Giraudat, J. (2000). Interactions between abscisic acid and ethylene signaling cascades. *The Plant Cell*, 12, 1103-1115.

[13] Bentsink, L, Jowett, J, Hanhart, C. J, & Koornneef, M. (2006). Cloning of *DOG1*, a quantitative trait locus controlling seed dormancy in Arabidopsis. *Proceedings of the National Academy of Sciences of the United States of America*, 103, 17042-17047.

[14] Bethke, P. C, Libourel, I. G. L, Aoyama, N, Chung, Y. Y, Still, D. W, & Jones, R. L. (2007). The Arabidopsis aleurone layer responds to nitric oxide, gibberellin, and abscisic acid and is sufficient and necessary for seed dormancy. *Plant Physiology*, 143, 1173-1188.

[15] Bewley, J. D. (1997). Seed germination and dormancy. *The Plant Cell*, , 9, 1055-1066.

[16] Bewley, J. D, & Black, M. (1994). *Seeds: Physiology of Development and Germination* (Plenum, New York).

[17] Blazquez, M. A, Ahn, J. H, & Weigel, D. (2003). A thermosensory pathway controlling flowering time in *Arabidopsis thaliana*. *Nature Genetics*, 33, 168-171.

[18] Bohlenius, H, Huang, T, Charbonnel-campaa, L, Brunner, A. M, Jansson, S, Strauss, S. H, & Nilsson, O. (2006). CO/FT regulatory module controls timing of flowering and seasonal growth cessation in trees. *Science*, 312, 1040-1043.

[19] Bond, D. M, & Finnegan, E. J. (2007). Passing the message on: inheritance of epigenetic traits. *Trends in Plant Science*, 12, 211-216.

[20] Bonnardeaux, Y, Li, C, Lance, R, Zhang, X, Sivasithamparam, K, & Appels, R. (2008). Seed dormancy in barley: identifying superior genotypes through incorporating epistatic interactions. *Australian Journal of Agricultural Research*, 59, 517-526.

[21] Borthwick, H. A, Hendricks, S. B, Parker, M. W, Toole, E. H, & Toole, V. K. (1952). A reversible photoreaction controlling seed germination. *Proceedings of the National Academy of Sciences of the United States of America*, 38, 662-666.

[22] Bothmer, R. V, & Jacobsen, N. (1985). Origin, taxonomy, and related species.D. C. Rasmusson, ed. Barley (American Society of Agronomists, Madison, Wisconsin, USA.), , 19-56.

[23] Boyko, A, Kathiria, P, Zemp, F. J, Yao, Y. L, Pogribny, I, & Kovalchuk, I. (2007). Transgenerational changes in the genome stability and methylation in pathogen-infected plants (Virus-induced plant genome instability). *Nucleic Acids Research*, 35, 1714-1725.

[24] Braybrook, S. A, Stone, S. L, Park, S, Bui, A. Q, Le, B. H, Fischer, R. L, Goldberg, R. B, & Harada, J. J. (2006). Genes directly regulated by LEAFY COTYLEDON2 provide insight into the control of embryo maturation and somatic embryogenesis. *Proceedings of the National Academy of Sciences of the United States of America*, 103, 3468-3473.

[25] Carles, C, Bies-etheve, N, Aspart, L, Leon-kloosterziel, K. M, Koornneef, M, Echeverria, M, & Delseny, M. (2002). Regulation of *Arabidopsis thaliana Em* genes: role of ABI5. *The Plant Journal*, 30, 373-383.

[26] Carrera, E, Holman, T, Medhurst, A, Peer, W, Schmuths, H, Footitt, S, Theodoulou, F. L, & Holdsworth, M. J. (2007). Gene expression profiling reveals defined functions of the ATP-binding cassette transporter COMATOSE late in phase II of germination. *Plant Physiology*, 143, 1669-1679.

[27] Casaretto, J. A, & Ho, T. H. (2005). Transcriptional regulation by abscisic acid in barley (*Hordeum vulgare* L.) seeds involves autoregulation of the transcription factor HvABI5. *Plant Molecular Biology*, 57, 21-34.

[28] Castro, A. J, Benitez, A, Hayes, P. M, Viega, L, & Wright, L. (2010). Coincident quantitative trait loci effects for dormancy, water sensitivity and malting quality traits in the BCD47 × Baronesse barley mapping population. *Crop and Pasture Science*, 61, 691-699.

[29] Chandler, P. M, Marion-poll, A, Ellis, M, & Gubler, F. (2002). Mutants at the Slender1 locus of barley cv Himalaya. molecular and physiological characterization. *Plant Physiology*, 129, 181-190.

[30] Chiang, G. C. K, Barua, D, Kramer, E. M, Amasino, R. M, & Donohue, K. (2009). Major flowering time gene, *FLOWERING LOCUS C*, regulates seed germination in *Arabidopsis thaliana*. *Proceedings of the National Academy of Sciences of the United States of America*, 106, 11661-11666.

[31] Chinnusamy, V, Gong, Z. Z, & Zhu, J. K. (2008). Abscisic acid-mediated epigenetic processes in plant development and stress responses. *Journal of Integrative Plant Biology*, 50, 1187-1195.

[32] Chiwocha, S. D, Cutler, A. J, Abrams, S. R, Ambrose, S. J, Yang, J, Ross, A. R, & Kermode, A. R. (2005). The *etr1-2* mutation in *Arabidopsis thaliana* affects the abscisic acid,

auxin, cytokinin and gibberellin metabolic pathways during maintenance of seed dormancy, moist-chilling and germination. *The Plant Journal*, 42, 35-48.

[33] Chouard, P. (1960). Vernalization and its relations to dormancy. *Annual Review of Plant Physiology and Plant Molecular Biology*, 11, 191-238.

[34] Corbineau, F, Picard, M. A, Fougereux, J. A, Ladonne, F, & Come, D. (2000). Effects of dehydration conditions on desiccation tolerance of developing pea seeds as related to oligosaccharide content and cell membrane properties. *Seed Science Research*, 10, 329-339.

[35] Crome, D, Lenoir, C, & Corbineau, F. (1984). The dormancy of cereals and its elimination. *Seed Science and Technology*, 12, 629-640.

[36] Debeaujon, I, Leon-kloosterziel, K. M, & Koornneef, M. (2000). Influence of the testa on seed dormancy, germination, and longevity in Arabidopsis. *Plant Physiology*, 122, 403-413.

[37] Derkx, M. P. M, & Karssen, C. M. (1993). Effects of light and temperature on seed dormancy and gibberellin-stimulated germination in *Arabidopsis thaliana*- studies with gibberellin-deficient and gibberellin-insensitive mutants. *Physiologia Plantarum*, 89, 360-368.

[38] Dill, A, Jung, H. S, & Sun, T. P. (2001). The DELLA motif is essential for gibberellin-induced degradation of RGA. *Proceedings of the National Academy of Sciences of the United States of America*, 98, 14162-14167.

[39] Eckardt, N. A. (2006). Genetic and epigenetic regulation of embryogenesis. *The Plant Cell*, 18, 781-784.

[40] Edney, M. J, & Mather, D. E. (2004). Quantitative trait loci affecting germination traits and malt friability in a two-rowed by six-rowed barley cross. *Journal of Cereal Science*, 39, 283-290.

[41] Ezcurra, I, Ellerstrom, M, Wycliffe, P, Stalberg, K, & Rask, L. (1999). Interaction between composite elements in the napA promoter: both the B-box ABA-responsive complex and the RY/G complex are necessary for seed-specific expression. *Plant Molecular Biology*, 40, 699-709.

[42] Finch-savage, W. E, & Leubner-metzger, G. (2006). Seed dormancy and the control of germination. *New Phytologist*, 171, 501-523.

[43] Finkelstein, R, Reeves, W, Ariizumi, T, & Steber, C. (2008). Molecular aspects of seed dormancy. *Annual Review of Plant Biology*, 59, 387-415.

[44] Finkelstein, R. R. (1994). Mutations at 2 new Arabidopsis ABA response loci are similar to the *abi3* mutations. *The Plant Journal*, 5, 765-771.

[45] Finkelstein, R. R, & Lynch, T. J. (2000). The arabidopsis abscisic acid response gene *ABI5* encodes a basic leucine zipper transcription factor. *The plant cell*, 12, 599-609.

[46] Finkelstein, R. R, Gampala, S. S. L, & Rock, C. D. (2002). Abscisic acid signaling in seeds and seedlings. *The Plant Cell*, 14, S15-S45.

[47] Finkelstein, R. R, Wang, M. L, Lynch, T. J, Rao, S, & Goodman, H. M. (1998). The Arabidopsis abscisic acid response locus *ABI4* encodes an APETALA2 domain protein. *The Plant Cell*, 10, 1043-1054.

[48] Flintham, J, Adlam, R, Bassoi, M, Holdsworth, M, & Gale, M. (2002). Mapping genes for resistance to sprouting damage in wheat. *Euphytica*, 126, 39-45.

[49] Flintham, J. E. (2000). Different genetic components control coat-imposed and embryo-imposed dormancy in wheat. *Seed Science Research*, 10, 43-50.

[50] Fu, X, Richards, D. E, Ait-ali, T, Hynes, L. W, Ougham, H, Peng, J, & Harberd, N. P. (2002). Gibberellin-mediated proteasome-dependent degradation of the barley DELLA protein SLN1 repressor. *The Plant Cell*, 14, 3191-3200.

[51] Fujii, H, Chinnusamy, V, Rodrigues, A, Rubio, S, Antoni, R, Park, S. Y, Cutler, S. R, Sheen, J, Rodriguez, P. L, & Zhu, J. K. (2009). *In vitro* reconstitution of an abscisic acid signalling pathway. *Nature*, 462, 660-664.

[52] Gagete, A. P, Riera, M, Franco, L, & Rodrigo, M. I. (2009). Functional analysis of the isoforms of an ABI3-like factor of *Pisum sativum* generated by alternative splicing. *Journal of Experimental Botany*, 60, 1703-1714.

[53] Gale, M. D, Flintham, J. E, & Devos, K. M. (2002). Cereal comparative genetics and preharvest sprouting. *Euphytica*, 126, 21-25.

[54] Gao, Y, Zeng, Q, Guo, J, Cheng, J, Ellis, B. E, & Chen, J. G. (2007). Genetic characterization reveals no role for the reported ABA receptor, GCR2, in ABA control of seed germination and early seedling development in Arabidopsis. *The Plant Journal*, 52, 1001-1013.

[55] Gao, W, Clancy, J. A, Han, F, Prada, D, Kleinhofs, A, & Ullrich, S. E. (2003). Molecular dissection of a dormancy QTL region near the chromosome 7 (5H) L telomere in barley. *Theoretical and Applied Genetics*, 107, 552-559.

[56] Gerjets, T, Scholefield, D, Foulkes, M. J, Lenton, J. R, & Holdsworth, M. J. (2009). An analysis of dormancy, ABA responsiveness, after-ripening and pre-harvest sprouting in hexaploid wheat (*Triticum aestivum* L.) caryopses. *Journal of Experimental Botany*, 61, 597-607.

[57] Ghassemian, M, Nambara, E, Cutler, S, Kawaide, H, Kamiya, Y, & Mccourt, P. (2000). Regulation of abscisic acid signaling by the ethylene response pathway in Arabidopsis. *The Plant Cell*, 12, 1117-1126.

[58] Gilroy, S, & Jones, R. L. (1994). Perception of gibberellin and abscisic-acid at the external face of the plasma-membrane of barley (*Hordeum vulgare* L) aleurone protoplasts. *Plant Physiology*, 104, 1185-1192.

[59] Giraudat, J, Hauge, B. M, Valon, C, Smalle, J, Parcy, F, & Goodman, H. M. (1992). Isolation of the arabidopsis-ABI3 gene by positional cloning. *The Plant Cell*, 4, 1251-1261.

[60] Gosti, F, Beaudoin, N, Serizet, C, Webb, A. A, Vartanian, N, & Giraudat, J. (1999). ABI1 protein phosphatase 2C is a negative regulator of abscisic acid signaling. *The Plant Cell*, 11, 1897-1910.

[61] Grappin, P, Bouinot, D, Sotta, B, Miginiac, E, & Jullien, M. (2000). Control of seed dormancy in *Nicotiana plumbaginifolia*: post-imbibition abscisic acid synthesis imposes dormancy maintenance. *Planta*, 210, 279-285.

[62] Griffiths, J, Murase, K, Rieu, I, Zentella, R, Zhang, Z. L, Powers, S. J, Gong, F, Phillips, A. L, Hedden, P, Sun, T. P, & Thomas, S. G. (2006). Genetic characterization and functional analysis of the GID1 gibberellin receptors in Arabidopsis. *The Plant Cell*, 18, 3399-3414.

[63] Groos, C, Gay, G, Perretant, M. R, Gervais, L, Bernard, M, Dedryver, F, & Charmet, D. (2002). Study of the relationship between pre-harvest sprouting and grain color by quantitative trait loci analysis in a whitexred grain bread-wheat cross. *Theoretical and Applied Genetics*, 104, 39-47.

[64] Gubler, F, Millar, A. A, & Jacobsen, J. V. (2005). Dormancy release, ABA and pre-harvest sprouting. *Current Opinion in Plant Biology*, 8, 183-187.

[65] Gubler, F, Kalla, R, Roberts, J. K, & Jacobsen, J. V. (1995). Gibberellin-regulated expression of a *MYB* gene in barley aleurone cells- evidence for *MYB* transactivation of a high-pI alpha-amylase gene promoter. *The Plant Cell*, 7, 1879-1891.

[66] Gubler, F, Chandler, P. M, White, R. G, Llewellyn, D. J, & Jacobsen, J. V. (2002). Gibberellin signaling in barley aleurone cells. Control of *SLN1* and *GAMYB* expression. *Plant Physiology*, 129, 191-200.

[67] Guo, J, Zeng, Q, Emami, M, Ellis, B. E, & Chen, J. G. (2008). The GCR2 gene family is not required for ABA control of seed germination and early seedling development in Arabidopsis. *PLoS ONE*, 3, e2982.

[68] Gutierrez, L, Van Wuytswinkel, O, Castelain, M, & Bellini, C. (2007). Combined networks regulating seed maturation. *Trends in Plant Science*, 12, 294-300.

[69] Gutterman, Y, Corbineau, F, & Come, D. (1996). Dormancy of *Hordeum spontaneum* caryopses from a population on the Negev Desert Highlands. *Journal of Arid Environments*, 33, 337-345.

[70] Han, F, Ullrich, S. E, Clancy, J. A, Jitkov, V, Kilian, A, & Romagosa, I. (1996). Verification of barley seed dormancy loci via linked molecular markers. *Theoretical and Applied Genetics*, 92, 87-91.

[71] Hattori, T, Terada, T, & Hamasuna, S. T. (1994). Sequence and functional analyses of the rice gene homologous to the maize Vp1. *Plant Molecular Biology*, 24, 805-810.

[72] Helliwell, C. A, Wood, C. C, Robertson, M, Peacock, W. J, & Dennis, E. S. (2006). The Arabidopsis FLC protein interacts directly *in vivo* with *SOC1* and *FT* chromatin and is part of a high-molecular-weight protein complex. *The Plant Journal*, 46, 183-192.

[73] Hickey, L. T, Lawson, W, Arief, V. N, Fox, G, Franckowiak, J, & Dieters, M. J. (2012). Grain dormancy QTL identified in a doubled haploid barley population derived from two non-dormant parents. *Euphytica*, 188, 113-122.

[74] Hilhorst, H. W. M. (1995). A critical update on seed dormancy.1. primary dormancy. *Seed Science Research*, 5, 61-73.

[75] Hoecker, U, Vasil, I. K, & Mccarty, D. R. (1995). Integrated control of seed maturation and germination programs by activator and repressor functions of Viviparous-1 of maize. *Genes & Development*, 9, 2459-2469.

[76] Holdsworth, M. J, Bentsink, L, & Soppe, W. J. J. (2008). Molecular networks regulating Arabidopsis seed maturation, after-ripening, dormancy and germination. *New Phytologist*, 179, 33-54.

[77] Horvath, D. (2009). Common mechanisms regulate flowering and dormancy. *Plant Science*, 177, 523-531.

[78] Huang, D. Q, Jaradat, M. R, Wu, W. R, Ambrose, S. J, Ross, A. R, Abrams, S. R, & Cutler, A. J. (2007). Structural analogs of ABA reveal novel features of ABA perception and signaling in Arabidopsis. *The Plant Journal*, 50, 414-428.

[79] Itoh, H, Ueguchi-tanaka, M, Sato, Y, Ashikari, M, & Matsuoka, M. (2002). The gibberellin signaling pathway is regulated by the appearance and disappearance of SLENDER RICE1 in nuclei. *The Plant Cell*, 14, 57-70.

[80] Jacobsen, J. V, Pearce, D. W, Poole, A. T, Pharis, R. P, & Mander, L. N. (2002). Abscisic acid, phaseic acid and gibberellin contents associated with dormancy and germination in barley. *Physiologia Plantarum*, 115, 428-441.

[81] Jang, Y. H, Lee, J. H, & Kim, J. K. (2008). Abscisic acid does not disrupt either the Arabidopsis FCA-FY interaction or its rice counterpart *in vitro*. *Plant and Cell Physiology*, 49, 1898-1901.

[82] Ji, H. S, Chu, S. H, Jiang, W. Z, Cho, Y. I, Hahn, J. H, Eun, M. Y, Mccouch, S. R, & Koh, H. J. (2006). Characterization and mapping of a shattering mutant in rice that corresponds to a block of domestication genes. *Genetics*, 173, 995-1005.

[83] Jiang, S, Kumar, S, Eu, Y. J, Jami, S. K, Stasolla, C, & Hill, R. D. (2012). The Arabidopsis mutant, fy-1, has an ABA-insensitive germination phenotype. *Journal of Experimental Botany*, 63, 2693-2703.

[84] Jones, H. D, Peters, N. C. B, & Holdsworth, M. J. (1997). Genotype and environment interact to central dormancy and differential expression of the VIVIPAROUS 1 homologue in embryos of *Avena fatua*. *The Plant Journal*, 12, 911-920.

[85] Kagaya, Y, Toyoshima, R, Okuda, R, Usui, H, Yamamoto, A, & Hattori, T. (2005). LEAFY COTYLEDON1 controls seed storage protein genes through its regulation of FUSCA3 and ABSCISIC ACID INSENSITIVE3. *Plant and Cell Physiology*, 46, 399-406.

[86] Kahn, A, Goss, J. A, & Smith, D. E. (1957). Effect of gibberellin on germination of lettuce seed. *Science*, 125, 645-646.

[87] Karssen, C. M, Swan, B. V, Breekland, A. E, & Koornneef, M. (1983). Induction of dormancy during seed development by endogenous abscisic acid: studies on abscisic acid deficient genotypes of *Arabidopsis thaliana* (L) *Heynh*. *Planta*, 157, 158-165.

[88] Kermode, A. R. (2005). Role of abscisic acid in seed dormancy. *Journal of Plant Growth Regulation*, 24, 319-344.

[89] Koornneef, M, Reuling, G, & Karssen, C. M. (1984). The isolation and characterization of abscisic acid insensitive mutants of *Arabidopsis thaliana*.. *Physiologia Plantarum*, 61, 377-383.

[90] Koornneef, M, Bentsink, L, & Hilhorst, H. (2002). Seed dormancy and germination. *Current Opinion in Plant Biology*, 5, 33-36.

[91] Koornneef, M, Hanhart, C. J, Hilhorst, H. W. M, & Karssen, C. M. (1989). *In vivo* inhibition of seed development and reserve protein accumulation in recombinants of abscisic-acid biosynthesis and responsiveness mutants in *Arabidopsis thaliana*. *Plant Physiology*, 90, 463-469.

[92] Kroj, T, Savino, G, Valon, C, Giraudat, J, & Parcy, F. (2003). Regulation of storage protein gene expression in Arabidopsis. *Development*, 130, 6065-6073.

[93] Kumar, S, Jiang, S, Jami, S. K, & Hill, R. D. (2011). Cloning and characterization of barley caryopsis FCA. *Physiologia Plantarum*, 143, 93-106.

[94] Lefebvre, V, North, H, Frey, A, Sotta, B, Seo, M, Okamoto, M, Nambara, E, & Marion-poll, A. (2006). Functional analysis of Arabidopsis NCED6 and NCED9 genes indicates that ABA synthesized in the endosperm is involved in the induction of seed dormancy. *The Plant Journal*, 45, 309-319.

[95] LePage-DegivryM.T., and Garello, G. ((1992). *In situ* abscisic acid synthesis : a requirement for induction of embryo dormancy in *Helianthus annuus*. *Plant Physiology*, 98, 1386-1390.

[96] Leubner-metzger, G. (2001). Brassinosteroids and gibberellins promote tobacco seed germination by distinct pathways. *Planta*, 213, 758-763.

[97] Leung, J, Merlot, S, & Giraudat, J. (1997). The Arabidopsis ABSCISIC ACID-INSENSITIVE2 (ABI2) and ABI1 genes encode homologous protein phosphatases 2C involved in abscisic acid signal transduction. *The Plant Cell*, 9, 759-771.

[98]   Leymarie, J, Bruneaux, E, Gibot-leclerc, S, & Corbineau, F. (2007). Identification of transcripts potentially involved in barley seed germination and dormancy using cDNA-AFLP. *Journal of Experimental Botany*, 58, 425-437.

[99]   Leymarie, J, Robayo-romero, M. E, Gendreau, E, Benech-arnold, R. L, & Corbineau, F. (2008). Involvement of ABA in induction of secondary dormancy in barley (*Hordeum vulgare* L.) seeds. *Plant and Cell Physiology*, 49, 1830-1838.

[100] Li, C. D, Tarr, A, Lance, R. C. M, Harasymow, S, Uhlmann, J, Westcot, S, Young, K. J, Grime, C. R, Cakir, M, Broughton, S, & Appelsa, R. (2003). A major QTL controlling seed dormancy and pre-harvest sprouting/grain alpha-amylase in two-rowed barley (*Hordeum vulgare* L.). *Australian Journal of Agricultural Research*, 54, 1303-1313.

[101] Li, B. L, & Foley, M. E. (1997). Genetic and molecular control of seed dormancy. *Trends in Plant Science*, 2, 384-389.

[102] Li, C, Ni, P, Francki, M, Hunter, A, Zhang, Y, Schibeci, D, et al. (2004). Genes controlling seed dormancy and pre-harvest sprouting in a rice-wheat-barley comparison. *Functional & Integrative Genomics*, 4, 84-93.

[103] Li, X, Qian, Q, Fu, Z, Wang, Y, Xiong, G, Zeng, D, Wang, X, Liu, X, Teng, S, Hiroshi, F, Yuan, M, Luo, D, Han, B, & Li, J. (2003). Control of tillering in rice. *Nature*, 422, 618-621.

[104] Lin, P. C, Hwang, S. G, Endo, A, Okamoto, M, Koshiba, T, & Cheng, W. H. (2007). Ectopic expression of abscisic acid 2/glucose insensitive 1 in arabidopsis promotes seed dormancy and stress tolerance. *Plant Physiology*, 143, 745-758.

[105] Liu, P. P, Montgomery, T. A, Fahlgren, N, Kasschau, K. D, Nonogaki, H, & Carrington, J. C. (2007a). Repression of auxin response factor10 by microrna160 is critical for seed germination and post-germination stages. *The Plant Journal*, 52, 133-146.

[106] Liu, X, Yue, Y, Li, B, Nie, Y, Li, W, Wu, W. H, & Ma, L. receptor is a plasma membrane receptor for the plant hormone abscisic acid. *Science*, 315, 1712-1716.

[107] Lohwasser, U, Roder, M. S, & Borner, A. (2005). QTL mapping of the domestication traits pre-harvest sprouting and dormancy in wheat (*Triticum aestivum* L.). *Euphytica*, 143, 247-249.

[108] Lopez-molina, L, Mongrand, B, Mclachlin, D. T, Chait, B. T, & Chua, N. H. (2002). ABI5 acts downstream of ABI3 to execute an ABA-dependent growth arrest during germination. *The Plant Journal*, 32, 317-328.

[109] Lotan, T, Ohto, M, Yee, K. M, West, M. A. L, Lo, R, Kwong, R. W, Yamagishi, K, Fischer, R. L, Goldberg, R. B, & Harada, J. J. (1998). Arabidopsis LEAFY COTYLEDON1 is sufficient to induce embryo development in vegetative cells. *Cell*, 93, 1195-1205.

[110] Luerssen, K, Kirik, V, Herrmann, P, & Misera, S. (1998). FUSCA3 encodes a protein with a conserved VP1/ABI3-like B3 domain which is of functional importance for the regulation of seed maturation in *Arabidopsis thaliana*. *The Plant Journal*, 15, 755-764.

[111] Oberthur, L, Blake, T. K, Dyer, W. E, & Ullrich, S. E. (1995). Genetic analysis of seed dormancy in barley (*Hordeum vulgare* L.). *Journal of Quantitative Trait Loci*.

[112] Ma, Y, Szostkiewicz, I, Korte, A, Moes, D, Yang, Y, Christmann, A, & Grill, E. (2009). Regulators of phosphatase activity function as abscisic acid sensors. *Science*pp. 1064-1068, 324, 2C.

[113] Mccarty, D. R, & Carson, C. B. (1991). The molecular-genetics of seed maturation in maize. *Physiologia Plantarum*, 81, 267-272.

[114] Mccarty, D. R, Carson, C. B, Stinard, P. S, & Robertson, D. S. (1989). Molecular analysis of Viviparous-1- an abscisic acid-insensitive mutant of maize. *The Plant Cell*, 1, 523-532.

[115] Mccarty, D. R, Hattori, T, Carson, C. B, Vasil, V, Lazar, M, & Vasil, I. K. (1991). The Viviparous-1 developmental gene of maize encodes a novel transcriptional activator. *Cell*, 66, 895-905.

[116] Mcginnis, K. M, Thomas, S. G, Soule, J. D, Strader, L. C, Zale, J. M, Sun, T. P, & Steber, C. M. (2003). The Arabidopsis SLEEPY1 gene encodes a putative F-box subunit of an SCF E3 ubiquitin ligase. *The Plant Cell*, 15, 1120-1130.

[117] Mckibbin, R. S, Wilkinson, M. D, Bailey, P. C, Flintham, J. E, Andrew, L. M, Lazzeri, P. A, Gale, M. D, Lenton, J. R, & Holdsworth, M. J. (2002). Transcripts of Vp-1 homeologues are misspliced in modern wheat and ancestral species. *Proceedings of the National Academy of Sciences of the United States of America*, 99, 10203-10208.

[118] Merlot, S, Gosti, F, Guerrier, D, Vavasseur, A, & Giraudat, J. (2001). The ABI1 and ABI2 protein phosphatases 2C act in a negative feedback regulatory loop of the abscisic acid signalling pathway. *The Plant Journal*, 25, 295-303.

[119] Meyer, K, Leube, M. P, & Grill, E. (1994). A Protein Phosphatase 2C involved in ABA signal-transduction in *Arabidopsis thaliana*. *Science*, 264, 1452-1455.

[120] Monke, G, Altschmied, L, Tewes, A, Reidt, W, Mock, H. P, Baumlein, H, & Conrad, U. (2004). Seed-specific transcription factors ABI3 and FUS3: molecular interaction with DNA. *Planta*, 219, 158-166.

[121] Muller, A. H, & Hansson, M. (2009). The barley magnesium chelatase 150-kd subunit is not an abscisic acid receptor. *Plant Physiology*, 150, 157-166.

[122] Nakaminami, K, Sawada, Y, Suzuki, M, Kenmoku, H, Kawaide, H, Mitsuhashi, W, Sassa, T, Inoue, Y, Kamiya, Y, & Toyomasu, T. (2003). Deactivation of gibberellin by 2-oxidation during germination of photoblastic lettuce seeds. *Bioscience Biotechnology Biochemistry*, 67, 1551-1558.

[123] Nakamura, S, Lynch, T. J, & Finkelstein, R. R. (2001). Physical interactions between ABA response loci of Arabidopsis. *The Plant Journal*, 26, 627-635.

[124] Nambara, E, & Marion-poll, A. (2003). ABA action and interactions in seeds. *Trends in Plant Science*, 8, 213-217.

[125] Nambara, E, Keith, K, Mccourt, P, & Naito, S. (1995). A regulatory role for the ABI3 gene in the establishment of embryo maturation in *Arabidopsis thaliana. Development*, 121, 629-636.

[126] Nambara, E, Hayama, R, Tsuchiya, Y, Nishimura, M, Kawaide, H, Kamiya, Y, & Naito, S. (2000). The role of *ABI3* and *FUS3* loci in *Arabidopsis thaliana* on phase transition from late embryo development to germination. *Developmental Biology*, 220, 412-423.

[127] Nikolaeva, M. G. (1969). Physiology of deep dormancy in seeds. National Science Foundation, Washington, DC, USA.

[128] Nishimura, N, Hitomi, K, Arvai, A. S, Rambo, R. P, Hitomi, C, Cutler, S. R, Schroeder, J. I, & Getzoff, E. D. (2009). Structural mechanism of abscisic acid binding and signaling by dimeric PYR1. *Science*, 326, 1373-1379.

[129] Noriyuki, N, Ali, S, Kazumasa, N, Sang-youl, P, Angela, W, Paulo, C. C, Stephen, L, Daniel, F. C, Sean, R. C, Joanne, C, John, R. Y, & Julian, I. S. (2010). PYR/PYL/RCAR family members are major *in vivo* ABI1 protein phosphatase 2C-interacting proteins in Arabidopsis. *The Plant Journal*, 61, 290-299.

[130] Ogas, J, Kaufmann, S, Henderson, J, & Somerville, C. (1999). PICKLE is a CHD3 chromatin-remodeling factor that regulates the transition from embryonic to vegetative development in Arabidopsis. *Proceedings of the National Academy of Sciences of the United States of America*, 96, 13839-13844.

[131] Ogawa, M, Hanada, A, Yamauchi, Y, Kuwalhara, A, Kamiya, Y, & Yamaguchi, S. (2003). Gibberellin biosynthesis and response during Arabidopsis seed germination. *The Plant Cell*, 15, 1591-1604.

[132] Oh, E, Yamaguchi, S, Kamiya, Y, Bae, G, Chung, W. I, & Choi, G. (2006). Light activates the degradation of PIL5 protein to promote seed germination through gibberellin in Arabidopsis. *The Plant Journal*, 47, 124-139.

[133] Oh, E, Yamaguchi, S, Hu, J. H, Yusuke, J, Jung, B, Paik, I, Lee, H. S, Sun, T. P, Kamiya, Y, & Choi, G. (2007). PIL5, a phytochrome-interacting bHLH protein, regulates gibberellin responsiveness by binding directly to the GAI and RGA promoters in Arabidopsis seeds. *The Plant Cell*, 19, 1192-1208.

[134] Okamoto, M, Kuwahara, A, Seo, M, Kushiro, T, Asami, T, Hirai, N, Kamiya, Y, Koshiba, T, & Nambara, E. which encode abscisic acid 8'-hydroxylases, are indispensable for proper control of seed dormancy and germination in Arabidopsis. *Plant Physiology*, 141, 97-107.

[135] Pandey, S, Nelson, D. C, & Assmann, S. M. (2009). Two novel GPCR-type G proteins are abscisic acid receptors in Arabidopsis. *Cell*, 136, 136-148.

[136] Parcy, F, Valon, C, Raynal, M, Gaubiercomella, P, Delseny, M, & Giraudat, J. (1994). Regulation of gene-expression programs during Arabidopsis seed development-roles of the *ABI3* locus and of endogenous abscisic-acid. *The Plant Cell*, 6, 1567-1582.

[137] Park, S. Y. (2009). Abscisic acid inhibits type 2C protein phosphatases via the PYR/PYL family of START proteins. *Science*, 324, 1068-1071.

[138] Penfield, S, Gilday, A. D, Halliday, K. J, & Graham, I. A. (2006). DELLA-mediated co-tyledon expansion breaks coat-imposed seed dormancy. *Current Biology*, 16, 2366-2370.

[139] Peng, J. R, & Harberd, N. P. (2002). The role of GA-mediated signalling in the control of seed germination. *Current Opinion in Plant Biology*, 5, 376-381.

[140] Piskurewicz, U, Tureckova, V, Lacombe, E, & Lopez-molina, L. (2009). Far-red light inhibits germination through DELLA-dependent stimulation of ABA synthesis and ABI3 activity. *Embo Journal*, 28, 2259-2271.

[141] Prada, D, Ullrich, S. E, Molina-cano, J. L, Cistué, L, Clancy, J. A, & Romagosa, I. (2004). Genetic control of dormancy in a Triumph/Morex cross in barley. *Theoretical and Applied Genetics, 109*, 62-70.

[142] Raz, V, Bergervoet, J. H. W, & Koornneef, M. (2001). Sequential steps for developmental arrest in Arabidopsis seeds. *Development*, 128, 243-252.

[143] Razem, F. A, El Kereamy, A, Abrams, S. R, & Hill, R. D. (2006). The RNA-binding protein FCA is an abscisic acid receptor. *Nature*, 439, 290-294.

[144] Razem, F. A, Luo, M, Liu, J. H, Abrams, S. R, & Hill, R. D. (2004). Purification and characterization of a barley aleurone abscisic acid-binding protein. *Journal of Biological Chemistry*, 279, 9922-9929.

[145] Reidt, W, Wohlfarth, T, Ellerstrom, M, Czihal, A, Tewes, A, Ezcurra, I, Rask, L, & Baumlein, H. (2000). Gene regulation during late embryogenesis: the RY motif of maturation-specific gene promoters is a direct target of the FUS3 gene product. *The Plant Journal, 21,* 401-408.

[146] Richards, D. E, Peng, J. R, & Harberd, N. P. (2000). Plant GRAS and metazoan STATs: one family. *Bioessays*, 22, 573-577.

[147] Rios, G, Gagetel, A. P, Castillo, J, Berbel, A, Franco, L, & Rodrigo, M. I. (2007). Abscisic acid and desiccation-dependent expression of a novel putative SNF5-type chromatin-remodeling gene in *Pisum sativum*. *Plant Physiology and Biochemistry*, 45, 427-435.

[148] Risk, J. M, Day, C. L, & Macknight, R. C. (2009). Reevaluation of abscisic acid-binding assays shows that G-Protein-Coupled Receptor2 does not bind abscisic acid. *Plant Physiology*, 150, 6-11.

[149] Robichaud, C, & Sussex, I. M. (1986). The response of viviparous-1 and wild-type embryos of *Zea mays* to culture in the presence of abscisic acid. *Journal of Plant Physiology*, 126, 235-242.

[150] Rodriguez, P. L, Leube, M. P, & Grill, E. (1998). Molecular cloning in *Arabidopsis thaliana* of a new protein phosphatase 2C (with homology to ABI1 and ABI2. *Plant Molecular Biology*pp. 879-883, 38, 2C.

[151] Rohde, A, Prinsen, E, De Rycke, R, Engler, G, Van Montagu, M, & Boerjan, W. (2002). PtABI3 impinges on the growth and differentiation of embryonic leaves during bud set in poplar. *The Plant Cell*, 14, 1885-1901.

[152] Ruttink, T, Arend, M, Morreel, K, Storme, V, Rombauts, S, Fromm, J, Bhalerao, R. P, Boerjan, W, & Rohde, A. (2007). A molecular timetable for apical bud formation and dormancy induction in poplar. *The Plant Cell*, 19, 2370-2390.

[153] Santiago, J, Dupeux, F, Round, A, Antoni, R, Park, S. Y, Jamin, M, Cutler, S. R, Rodriguez, P. L, & Marquez, J. A. (2009). The abscisic acid receptor PYR1 in complex with abscisic acid. *Nature*, 462, 665-668.

[154] Sasaki, A, Itoh, H, Gomi, K, Ueguchi-tanaka, M, Ishiyama, K, Kobayashi, M, Jeong, D. H, An, G, Kitano, H, Ashikari, M, & Matsuoka, M. (2003). Accumulation of phosphorylated repressor for gibberellin signaling in an F-box mutant. *Science*, 299, 1896-1898.

[155] Sawada, Y, Aoki, M, Nakaminami, K, Mitsuhashi, W, Tatematsu, K, Kushiro, T, Koshiba, T, Kamiya, Y, Inoue, Y, Nambara, E, & Toyomasu, T. (2008). Phytochrome- and gibberellin-mediated regulation of abscisic acid metabolism during germination of photoblastic lettuce seeds. *Plant Physiology*, 146, 1386-1396.

[156] Seo, M, Hanada, A, Kuwahara, A, Endo, A, Okamoto, M, Yamauchi, Y, North, H, Marion-poll, A, Sun, T. P, Koshiba, T, Kamiya, Y, Yamaguchi, S, & Nambara, E. (2006). Regulation of hormone metabolism in Arabidopsis seeds: phytochrome regulation of abscisic acid metabolism and abscisic acid regulation of gibberellin metabolism. *The Plant Journal*, 48, 354-366.

[157] Sheldon, C. C, Rouse, D. T, Finnegan, E. J, Peacock, W. J, & Dennis, E. S. (2000). The molecular basis of vernalization: The central role of FLOWERING LOCUS C (FLC). *Proceedings of the National Academy of Sciences of the United States of America*, 97, 3753-3758.

[158] Shen, Y. Y, Wang, X. F, Wu, F. Q, Du, S. Y, Cao, Z, Shang, Y, Wang, X. L, Peng, C. C, Yu, X. C, Zhu, S. Y, Fan, R. C, Xu, Y. H, & Zhang, D. P. (2006). The Mg-chelatase H subunit is an abscisic acid receptor. *Nature*, 443, 823-826.

[159] Stone, S. L, Kwong, L. W, Yee, K. M, Pelletier, J, Lepiniec, L, Fischer, R. L, Goldberg, R. B, & Harada, J. J. (2001). LEAFY COTYLEDON2 encodes a B3 domain transcription factor that induces embryo development. *Proceedings of the National Academy of Sciences of the United States of America*, 98, 11806-11811.

[160] Suzuki, M, Kao, C. Y, Cocciolone, S, & Mccarty, D. R. complements Arabidopsis abi3 and confers a novel ABA/auxin interaction in roots. *The Plant Journal*, 28, 409-418.

[161] Suzuki, T, Matsuura, T, Kawakami, N, & Noda, K. (2000). Accumulation and leakage of abscisic acid during embryo development and seed dormancy in wheat. *Plant Growth Regulation*, 30, 253-260.

[162] Sweeney, M. T, Thomson, M. J, Pfeil, B. E, & Mccouch, S. (2006). Caught red-handed: Rc encodes a basic helix-loop-helix protein conditioning red pericarp in rice. *The Plant Cell*, 18, 283-294.

[163] Taylor, M, Boland, M, & Brester, G. (2009). Barley Profile (AgMRC, USDA).

[164] Tiedemann, J, Rutten, T, Monke, G, Vorwieger, A, Rolletschek, H, Meissner, D, Milkowski, C, Petereck, S, Mock, H. P, Zank, T, & Baumlein, H. (2008). Dissection of a complex seed phenotype: Novel insights of FUSCA3 regulated developmental processes. *Developmental Biology*, 317, 1-12.

[165] To, A, Valon, C, Savino, G, Guilleminot, J, Devic, M, Giraudat, J, & Parcy, F. (2006). A network of local and redundant gene regulation governs Arabidopsis seed maturation. *The Plant Cell*, 18, 1642-1651.

[166] Toledo-ortiz, G, Huq, E, & Quail, P. H. (2003). The Arabidopsis basic/helix-loop-helix transcription factor family. *The Plant Cell*, 15, 1749-1770.

[167] Toyomasu, T, Kawaide, H, Mitsuhashi, W, Inoue, Y, & Kamiya, Y. (1998). Phytochrome regulates gibberellin biosynthesis during germination of photoblastic lettuce seeds. *Plant Physiology*, 118, 1517-1523.

[168] Ueguchi-tanaka, M, Ashikari, M, Nakajima, M, Itoh, H, Katoh, E, Kobayashi, M, Chow, T. Y, Hsing, Y. I. C, Kitano, H, Yamaguchi, I, & Matsuoka, M. (2005). Gibberellin insensitive dwarf1 encodes a soluble receptor for gibberellin. *Nature*, 437, 693-698.

[169] Ueguchi-tanaka, M, Nakajima, M, Katoh, E, Ohmiya, H, Asano, K, Saji, S, Xiang, H. Y, Ashikari, M, Kitano, H, Yamaguchi, I, & Matsuokaa, M. (2007). Molecular interactions of a soluble gibberellin receptor, GID1, with a rice DELLA protein, SLR1, and gibberellin. *The Plant Cell*, 19, 2140-2155.

[170] Ullrich, S. E, Hays, P. M, Dyer, W. E, Black, T. K, & Clancy, J. A. (1993). Quantitative trait locus analysis of seed dormancy in Steptoe barley. In: Walker-Simmons MK, Ried JL (eds) Preharvest sprouting in cereals 1992. American Association of Cereal Chemistry, St Paul, , 136-145.

[171] Ullrich, S. E, Han, F, Gao, W, Prada, D, Clancy, J. A, Kleinhofs, A, Romagosa, I, & Molina-cano, J. L. (2002). Summary of QTL analyses of the seed dormancy trait in

barley. Barley Newsletter Available at: http://wheat.pw.usda.gov/ggpages/Barley-Newsletter/45/Proceedings1.html, 45, 39-41.

[172]   Ullrich, S. E, Han, F, Blake, T. K, Oberthur, L. E, Dyer, W. E, & Clancy, J. A. (1995). Seed dormancy in barley: genetic resolution and relationship to other traits. In: Noda K, Mares DJ, editors. *Pre-harvest sprouting in cereals.* Osaka: Center for Academic Societies Japan; 1996. , 157-163.

[173]   Ullrich, S. E, Lee, H, & Clancy, J. A. del Blanco, I.A., Jitkov, V.A., Kleinhofs, A., Han, F., Prada, D., Romagosa, I., and Molina-Cano, J.L. ((2009). Genetic relationships between preharvest sprouting and dormancy in barley. *Euphytica*, 168, 331-345.

[174]   Verslues, P. E, & Zhu, J. K. (2007). New developments in abscisic acid perception and metabolism. *Current Opinion in Plant Biology*, 10, 447-452.

[175]   Walck, J. L, Baskin, J. M, Baskin, C. C, & Hidayati, S. N. (2005). Defining transient and persistent seed banks in species with pronounced seasonal dormancy and germination patterns. *Seed Science Research*, 15, 189-196.

[176]   Walker-simmons, M. and sensitivity in developing wheat embryos of sprouting resistant and susceptible cultivars. *Plant Physiology*, 84, 61-66.

[177]   Wasilewska, A, Vlad, F, Sirichandra, C, Redko, Y, Jammes, F, Valon, C, Frey, N. F. D, & Leung, J. (2008). An update on abscisic acid signaling in plants and more. *Molecular Plant*, 1, 198-217.

[178]   Wen, C. K, & Chang, C. (2002). Arabidopsis *RGL1* encodes a negative regulator of gibberellin responses. *The Plant Cell*, 14, 87-100.

[179]   Wilkinson, M, Lenton, J, & Holdsworth, M. (2005). Transcripts of *VP-1* homoeologues are alternatively spliced within the *Triticeae* tribe. *Euphytica*, 143, 243-246.

[180]   Willige, B. C, Ghosh, S, Nill, C, Zourelidou, M, Dohmann, E. M. N, Maier, A, & Schwechheimer, C. (2007). The DELLA domain of GA INSENSITIVE mediates the interaction with the GA INSENSITIVE DWARF1A gibberellin receptor of Arabidopsis. *The Plant Cell*, 19, 1209-1220.

[181]   Wu, F. Q, Xin, Q, Cao, Z, Liu, Z. Q, Du, S. Y, Mei, C, Zhao, C. X, Wang, X. F, Shang, Y, Jiang, T, Zhang, X. F, Yan, L, Zhao, R, Cui, Z. N, Liu, R, Sun, H. L, Yang, X. L, Su, Z, & Zhang, D. P. (2009). The magnesium-chelatase H subunit binds abscisic acid and functions in abscisic acid signaling: new evidence in arabidopsis. *Plant Physiology*, 150, 1940-1954.

[182]   Wu, K. Q, Tian, L. N, Zhou, C. H, Brown, D, & Miki, B. (2003). Repression of gene expression by Arabidopsis HD2 histone deacetylases. *The Plant Journal*, 34, 241-247.

[183]   Yamashino, T, Matsushika, A, Fujimori, T, Sato, S, Kato, T, Tabata, S, & Mizuno, T. (2003). A link between circadian-controlled bHLH factors and the APRR1/TOC1 quintet in Arabidopsis thaliana. *Plant and Cell Physiology*, 44, 619-629.

[184] Yamauchi, Y, Takeda-kamiya, N, Hanada, A, Ogawa, M, Kuwahara, A, Seo, M, Kamiya, Y, & Yamaguchi, S. (2007). Contribution of gibberellin deactivation by At-GA2ox2 to the suppression of germination of dark-imbibed *Arabidopsis thaliana* seeds. *Plant and Cell Physiology*, 48, 555-561.

[185] Yang, J. H, Seo, H. H, Han, S. J, Yoon, E. K, Yang, M. S, & Lee, W. S. (2008). Phytohormone abscisic acid control RNA-dependent RNA polymerase 6 gene expression and post-transcriptional gene silencing in rice cells. *Nucleic Acids Research*, 36, 1220-1226.

[186] Yanovsky, M. J, & Kay, S. A. (2002). Molecular basis of seasonal time measurement in Arabidopsis. *Nature*, 419, 308-312.

[187] Zhang, D. P, Wu, Z. Y, Li, X. Y, & Zhao, Z. X. (2002). Purification and identification of a 42-kilodalton abscisic acid-specific-binding protein from epidermis of broad bean leaves. *Plant Physiology*, 128, 714-725.

[188] Zhang, F, Chen, G, Huang, Q, Orion, O, Krugman, T, Fahima, T, et al. (2005). Genetic basis of barley caryopsis dormancy and seedling desiccation tolerance at the germination stage. *Theoretical and Applied Genetics*, 110, 445-453.

[189] Zohary, D, & Hopf, M. (1993). Domestication of plants in the Old World. The origin and spread of cultivated plants in West Asia, Europe and the Nile Valley. *Clarendon Press, Oxford, England.*

[190] Zou, M, Guan, Y, Ren, H, Zhang, F, & Chen, F. (2007). Characterization of alternative splicing products of bZIP transcription factors OsABI5. Biochemical and Biophysical Research Communications, 360, 307-313.

Chapter 4

Evaluation of Pigeonpea Germplasm for Important Agronomic Traits in Southern Africa

E. T. Gwata and H. Shimelis

Additional information is available at the end of the chapter

1. Introduction

Pigeonpea [*Cajanus cajan* (L.) Millsp.] is an important grain legume that originated in the Indian sub-continent. It is now grown in many parts of the world including southern Africa particularly the region encompassing Kenya, Mozambique, Malawi and southern Tanzania (Høgh-Jensen *et al.*, 2007). This region is considered as a secondary centre of diversity for pigeonpea. The diversity associated with the pigeonpea germplasm from the region was documented widely (Songok *et al.*, 2010; Mligo and Craufurd 2005; Silim *et al.*, 2005).

The crop is grown for its multiple benefits mainly by smallholder growers and is useful in providing household food security in the region. The crop provides highly nutritious food for human consumption (Amarteifio *et al.*, 2002) and fixes considerable amounts of atmospheric nitrogen (Mapfumo *et al.*, 1999), thus improving soil fertility. Considerable quantities of the grain are traded within the region and in international markets particularly in the Indian sub-continent thus generating income for farmers. In addition, the stover is used for fuelwood and building material in some of the rural communities in the region (Silim *et al.*, 2005). Pigeonpea is also useful for controlling soil erosion in those areas prone to floods. The crop is also relatively tolerant to drought (Kumar *et al.*, 2011) thus making it suitable for cultivation in the semi-arid agro-ecological conditions prevalent in the region.

However, the average grain yield obtained by farmers in the region is generally low. In Tanzania, growers obtained 0.4 t/ha (Mligo and Myaka, 1994). This is partly because some of the smallholder pigeonpea growers cultivate largely unimproved landraces in mixed cropping systems (Fig. 1) partly because of the pressure of limited land for cultivating crops as well as the need to minimize the risk of crop failure. In addition, typical smallholder farmers are subsistent. Therefore, the broad objective of the study reported in this Chapter was to evaluate improved pigeonpea germplasm for agronomic performance across the region under rain-fed field conditions. The specific objectives were to evaluate the germplasm for (i) sensitivity to photoperiod (ii) reaction to fusarium wilt (iii) reaction to insect pests and (iv) grain quality traits that are preferred by end-users.

Figure 1. A mixed cropping system consisting of pigeonpea (foreground), corn, sorghum and cowpea.

2. Pigeonpea types

A significant proportion of the smallholder farmers in the region largely grows traditional landraces. The landraces are characterized by late maturity, inherently low grain yield and dark seeds. However, the landraces are adapted to the local biotic and abiotic stresses. In particular, they are tolerant to severe droughts that occur in the region. On the other hand, the improved cultivars fall into either short-duration (SD) or medium-duration (MD) or long-duration (LD) types. This classification is based on the duration to maturity.

The short-duration types require about 90 days in order to mature. Therefore, they mature in the middle of the rainy season (in the region) when post-harvest handling is difficult. This renders the grain susceptible to spoilage by fungal diseases in particular. Consequently, this type of pigeonpea is poorly preferred by farmers in the region. On the other hand, medium duration (MD) types require about 150 days in order to attain maturity while long-duration (LD) types can require up to 240 days to mature fully. The majority of the landraces in the region fall into this category. In general, late maturity in pigeonpea is attributed to sensitivity to day length (or photoperiod).

3. Production limitations

The production of pigeonpea in southern Africa is constrained by a range of abiotic and biotic factors. In particular, the crop is sensitive to photoperiod. The crop is also threatened by fusarium wilt (Gwata et al., 2006; Kannaiyan et al., 1989) and a broad range of insect pests (Minja, 1997; Minja et al., 1996).

3.1. Sensitivity to photoperiod (day length)

When the crop is grown in high latitude areas (>10° away from the equator), it is sensitive to photoperiod and temperature (Silim et al., 2006) with plant height, vegetative biomass, phenology and grain yield being affected most (Whiteman et al., 1985). Consequently, the delayed flowering and maturity lead to increased susceptibility to terminal drought that frequently occurs in southern Africa. Therefore, the cultivation of the late maturing LD types in the region pauses many challenges for the smallholder farmers. For instance, the winter season (which commences in June in the region) is associated with frost and generally low temperatures, to which pigeonpea is susceptible. Furthermore, after harvesting the main crops (during May), the small-holder farmers traditionally release their domestic livestock to graze freely (or unattended) in the fields. Such livestock interfere with late maturing crops that may still be growing in the fields. In addition, the delay in crop maturity may interfere with the timing of the succeeding crop. Therefore, this makes it difficult for farmers to develop consistent crop management practices and predictable cropping systems. In terms of marketing, the pigeonpea grain from the region is exported mainly to international markets in the Indian sub-continent where the prices are attractive before the glut in November. Therefore, the pigeonpea growers in southern Africa require pigeonpea types that can flower and mature early in order to have ample time for processing the grain for export to these distant markets when demand is at a peak.

3.2. Susceptibility to fusarium wilt

Apart from sensitivity to photoperiod, pigeonpea is threatened by the fusarium wilt disease caused by the fungal pathogen *Fusarium udum* Butler (Kannaiyan et al., 1989). It is the most devastating disease of pigeonpea in the region. The pathogen lives in the soil. Between crops, it survives in residual plant debris as mycelium and in all its spore forms (Agrios, 1997). The germ tube of the mycelium or spore penetrates seedlings through root tips, wounds or point of formation of lateral roots. The mycelium advances through the xylem causing vascular plugging followed by wilting of stems during flowering and pod-filling stages thus causing yield loss ranging from 30 to 100% (Reddy et al., 1990). Once a field is infested, the pathogen may survive in the soil for several years. The fungal spores can be disseminated to new plants by farm equipment, water, wind or animals, including humans. The use of cultivars resistant to the fungus is the most effective measure for controlling the disease.

3.3. Susceptibility to insect pests

Pigeonpea is susceptible to a wide range of insect pests that attack the crop at both the vegetative and reproductive stages (Minja *et al.*, 1999). Among the pests, the pod borer is regarded as a major threat to pigeonpea because of its destructiveness and extensive host range while pod sucking bugs and thrips can cause up to 78% (Dialoke *et al.*, 2010) and 47% (Rotimi and Iloba 2008) yield loss respectively.

Currently, the production area of this pigeonpea is expanding to non-traditional areas such as the semi-arid belt of the Limpopo River Basin (LRB) in southern Africa. However, the occurrence of insect species of economic importance in pigeonpea has not been investigated in the LRB.

3.4. End-use qualities

In southern Africa, pigeonpea is consumed mainly as whole fresh green peas. Usually, these fresh beans are boiled after or before shelling. In general, end-users in the region prefer large (100-grain weight = 15.0 ± 2.0 g). Where the dry peas are utilized for human consumption, the end-users prefer the large white (cream) bold types that are easy to cook. In contrast, landraces originating from central Africa possess small hard seeds which have no commercial value in the regional markets since end-users prefer large-seeded types that are easier to cook. Favourable end-use qualities also influence cultivar adoption by growers. Therefore, grain color and size measurements were evaluated as integral components of the field studies reported in this Chapter.

4. Field evaluation studies

Pigeonpea germplasm was evaluated under rain-fed field conditions across the region. The specific genotypes were selected on the basis of preliminary information obtained from previous large-scale screening of many pigeonpea genotypes in the field (Mogashoa and Gwata 2009), seed availability as well as local farmer-preferences in the area represented by each testing location.

4.1. Evaluation for photoperiod sensitivity

The major objective of this study was to evaluate pigeonpea germplasm that was developed previously for production in high latitude areas (>10° away from the equator) for adaptation as measured by the agronomic performance. The evaluation was conducted under rain-fed conditions initially at Chitedze (Malawi; 14° S) and subsequently at Thohoyandou (South Africa, 22 °S) testing locations.

4.1.1. Field evaluation at Chitedze

Six elite genotypes and two check cultivars were used in the study conducted at Chitedze. The experiment was arranged as a randomized complete block design replicated three times. At the beginning of the cropping season (in early December), seed of each genotype was sown manually in field plots, each measuring 5.0 m in length and containing five rows spaced at 1.2 m apart with 0.5 m between plants in the row. Standard agronomic management recommendations for pigeonpea were followed throughout the season. In each season, no chemical fertilizers were applied on the crop in consistency with other similar studies (Silim *et al.*, 2006). In particular, inorganic N fertilizer was deemed unnecessary for the crop since pigeonpea can symbiotically fix about 40-160 kg/ha of N per season (Mapfumo *et al.*, 1999). Pigeonpea is also able to access forms of phosphorus that are normally poorly available in the soil (Ae *et al.*, 1990). This is achieved through the presence of piscidic acid exudates that solubilize phosphorus in the rhizosphere (Ae *et al.*, 1990).

During the evaluation, four key indicators for agronomic performance namely the number of days to 50% flowering (50% DF), the number of days to 75% physiological maturity (75% DM), grain size as measured by 100-grain weight (100-GW) and grain yield were measured (Table 1). Statistical analysis of data sets using statistical analysis system (SAS) procedures (SAS Institute, 1989) was applied. Tukey's method (Ott, 1988) was applied to separate the trait means obtained for each respective set of the five genotypes.

Cultivar Code	Mean				
	50% DF (d)	75% DM (d)	Grain color	100-Grain Weght (g)	Grain Yield (t/ha)
01144/13	86 b	115 c	White	13.7 a	2.7 a
01160/15	112 a	167 a	White	14.4 a	2.4 a
01480/32	102 ab	166 a	White	13.8 a	3.0 a
01162/21	102 ab	163 ab	White	14.9 a	2.6 a
01167/11	96 ab	166 a	White	15.5 a	2.2 a
01514/15	84 b	153 bc	White	14.4 a	2.9 a
Mean	97	161	-	14.0	2.7
Royes*	83 b	173 a	White	13.7 a	1.0 b
MtawaJuni**	119 a	172 a	Brown	16.8 a	1.1 b

Means in the same column followed by the same letter are not significantly different at the 0.05 probability level by Tukey's test. *Commercial cultivar in Malawi; **Unimproved traditional landrace popular in Malawi.

Source: Adapted from Gwata and Siambi, 2009.

Table 1. Agronomic performance of pigeonpea germplasm evaluated under field conditions at Chitedze (Malawi, 14°S).

The newly developed germplasm showed considerable improvement in terms of duration to flowering and maturity as well as yield potential. For instance, in order to attain 50% flowering, cultivar '01144/13' and the unimproved local landrace MtawaJuni required 86 d and 119 d respectively (Table 1). Cultivar '01514/15' matured significantly ($P < 0.01$) earlier (153 d) than both the commercial cultivar Royes (173 d) and the landrace MtawaJuni (172 d). On average, the time to flowering and maturity among the new germplasm was reduced by 10 d in comparison with that for the unimproved landrace. The highest grain yield (3.0 t/ha) was observed for the cultivar '01480/32' compared with 1.0 t/ha that was observed for the check cultivar 'Royes' (Table 1). This indicated a three-fold higher potential for the new pigeonpea technologies in the region.

4.1.2. Field evaluation at Thohoyandou

The germplasm was subsequently introduced to non-traditional areas in the semi-arid LRB (Fig. 2) as represented by the testing location at Thohoyandou (Limpopo, South Africa) which is a typical ecotope representing the agro-ecological conditions in the region (Mzezewa et al., 2010). The soils at the location are predominantly deep (>150 cm), red and well drained clays with an apedal structure. The clay content is generally high (60 %) and soil reaction is acidic (pH 5.0).

The daily temperatures at the location vary from about 25°C to 40°C in summer and between approximately 12°C and 26°C in winter. Rainfall is highly seasonal with 95% occurring between October and March, often with a mid-season dry spell during critical periods of crop growth (FAO, 2009). Mid-season drought often leads to crop failure and low yields (Beukes et al., 1999). This spatial and temporal variability in annual rainfall experienced in the area imposes several major challenges for smallholder farmers mainly because their crop choices must take into consideration the challenges imposed by moisture stress during the cropping season. In this regard, the use of drought tolerant crops such as pigeonpea as a means for achieving sustainable crop production in the area is merited.

Nineteen exotic MD genotypes of pigeonpea (obtained from the International Crops Research Institute for the Semi-Arid Tropics, Nairobi, Kenya) as well as one unimproved landrace obtained from Limpopo [designated Limpopo Local (LL)] were used in the study. The evaluation was conducted over two cropping seasons (2008/2009 and 2009/2010) at Thohoyandou (596 m a.s.l.; 22º 58'S, 30º 26'E) in Limpopo Province (South Africa) following the method described above (Section 4.1.1). In each season, the experiment was laid out as a 5x4 lattice design replicated three times. At physiological maturity, yield attributes and the grain yield were measured.

The results showed that at least five cultivars produced >1.5 t/ha with cultivar '01508/10' obtaining the highest (2.36 t/ha) grain yield at this testing location. Apart from the local check cultivar, 30% of the cultivars evaluated in the field trial produced <0.5 t/ha indicating that they were low yielding under the agro-ecological conditions at Thohoyandou (Limpopo). However, the average grain yield obtained from the trial was about 1.01 t/ha. The average grain yield among the best five performing cultivars (1.98 t/ha) represented at least 70.0% more productivity relative to the trial mean. In comparison with the local check, cultivar '01508/10' produced almost ten-fold more grain yield indicating the potential increase in pigeonpea productivity in the area. In contrast, cultivar '01480/32' produced a low (< 0.6 t/ha) grain yield compared to 3.0 t/ha that was observed for this cultivar at Chitedze. This was expected since grain yield is a quantitative trait that is influenced by the environment. Subsequently, partly because of their high yield potential, appropriate time to maturity and good grain attributes, these cultivars were adopted widely by growers in the region.

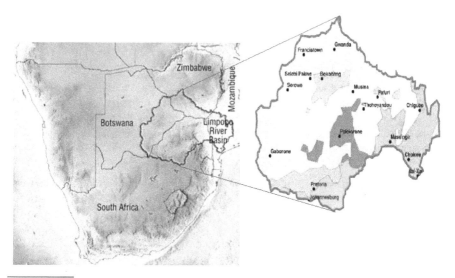

(*Source*: http://www.fao.org/docrep/008/y5744e/y5744e05.htm. Accessed 16 November, 2012).

Figure 2. The Limpopo River Basin in southern Africa.

4.2. Evaluation for reaction to fusarium wilt

The objective of this study was to evaluate selected pigeonpea germplasm under high disease pressure across the southern Africa region in order to identify resistant cultivars that produce optimum yields under high disease pressure.

Selected local and exotic pigeonpea genotypes were evaluated in three countries in eastern and southern Africa (Table 2). The genotypes used at each testing location were selected on the basis of preliminary information obtained from previous large-scale screening of many pigeonpea genotypes in the field, seed availability as well as local farmer-preferences in the area. The evaluation of the germplasm was conducted in wilt-sick plots (Bayaa *et al.*, 1997) at testing locations where the disease pressure was considered to be high. At the beginning of the cropping season, seed of each genotype was sown in field plots using the same method as described above (see section 4.1.1).

At physiological maturity, the percent incidence of fusarium wilt (% FW) was determined. Initially, individual plants in each plot were scored for wilting (as a symptom of *F. udum*) followed by visual examination of the cross-section of the stem of each candidate plant in order to confirm the presence of a brown ring of discoloured xylem vessels. During harvesting, grain size as measured by 100-grain weight (100-GW) and grain yield were measured. Data sets were analyzed using standard analysis of variance procedures followed by mean separation as described above (see section 4.1.1).

Both the highest (92.0) and lowest (1.7) % FW scores were observed at Ngabu (Malawi) for '00068' and '00020', respectively (Table 2). The disease incidence in the local cultivar (Royes) was 90.2% compared with <5.0% for cultivar '00040'. However, the disease incidence in '00040' was consistently low (<20.0%) at all three locations. Grain yield was influenced by the location. Nevertheless, at least 1.5 t/ha of grain yield was obtained for '00040' compared with <1.0 t/ha for the susceptible genotype (Table 2). At Ilonga testing location, the unimproved traditional landrace (Ex-Loguba-1) attained a low grain yield (1.3 t/ha) as well as size (100-GW = 10.1 g) but the elite genotypes ('00020' and '00040') averaged 2.7 t/ha.

Cultivar Code	Type	Mean Percent Fusarium Wilt (%)	Mean Grain Yield (t/ha)		
			Kenya (Kiboko)	Malawi (Ngabu)	Tanzania (Ilonga)
00040	Improved, resistant	12.4	2.2 c	1.9 b	3.0 b
00020	Improved, moderately resistant	13.9	1.6 b	1.7 b	2.4 b
00068 (Check)	Improved, susceptible	90.1	0.6 a	0.1 a	0.1 a
Royes (Check)	Improved, susceptible	-	-	0.6 a	-
Ex-Loguba-1 (Check)	Unimproved, landrace	-	-	-	1.3 b

Means in the same column followed by the same letter are not significantly different at the 0.05 probability level by Tukey's test.

Source: Gwata *et al.,* 2006

Table 2. Performance of improved pigeonpea cultivars under high disease pressure in eastern and southern Africa.

The results of this study indicated a high level of wilt resistance in the elite pigeonpea germplasm particularly '00040'. The classification of this cultivar as resistant to fusarium wilt was consistent with approaches used in classifying resistance to the disease in other leguminous species. For instance, genotypes showing <10% (Halila and Strange, 1997) and <20% (Bayaa *et al.,* 1997) incidence of fusarium wilt were considered resistant in chickpea (*Cicer arietinum*) and lentil (*Lens culinaris*), respectively. Because of its ability to withstand the high disease pressure, wide adaptability and high yield potential, '00040' was adopted widely by pigeonpea farmers in the region. Pigeonpea genotypes often exhibit differential host responses to *F. udum* suggesting that probably, at least two different pathogenic races of the disease exist in the region. In chickpea, Tekeoglu *et al.* (2000) reported lines with resistance to one race of fusarium wilt but susceptible to another race, which suggested that different resistance genes confer resistance to different races. Such lines could be useful as race differentials to facilitate identification of races based on host pathogen interactions. The disease resistance observed in this study could be useful as a good source of resistance in pigeonpea breeding programmes in the region.

4.3. Evaluation for insect pests

The objective of this study was to identify insect species that occur in pigeonpea and damage the crop particularly at the reproductive growth stage in the LRB. This information would be useful in designing more detailed investigations on the economic impact of such species on the crop. Twenty exotic pigeonpea genotypes were planted in a field experiment at Thohoyan-dou following the procedure described above (see section 4.1.1). In order to achieve optimum natural infestation, no pesticides or control measures were applied on the crop throughout the season.

Starting at the flower initiation stage through to 50% physiological maturity of the crop, random samples of above-ground insects were collected using the active insect sampling approach that utilizes a combination of scouting and the sweep net technique which has been used successfully in a variety of crops including soybean and cotton (Marston *et al.*, 1976; Spurgeon and Cooper, 2011). Sampling was performed weekly between 14.00 - 15.00 h when insect activity was at its peak. Sampled insects were collected into a specimen jar (15.0 cm in diameter x 25.0 cm) containing ethyl acetate. Individual specimens were identified to the species level in the laboratory using a pigeonpea pest identification handbook (Dialoke *et al.*, 2010; Night and Ogenga-Latigo, 1994). Identification to the species level is necessary since there is a possibility that species within a family and even genera can exhibit differential host plants and natural enemy complexes.

The study showed that a broad range of insect species injurious to pigeonpea was collected from the crop and identified (Table 3). The species included those that are known to be pests in pigeonpea such as blister beetles, pod suckers, flower thrips and the pod borer which is regarded as a serious threat to pigeonpea because of its destructiveness and extensive host range. The range of insects found in the study was in agreement with that reported from the surveys conducted in other parts of Africa confirming that a fairly common spectrum of insect pests occurs in the crop across the African sub-continent (Dialoke *et al.*, 2010; Minja *et al.*, 1999). In addition, most of the insect species identified in this study have been classified as pests of pigeonpea elsewhere (Minja, 2001; Shanower *et al.*, 1999; Rotimi and Iloba 2008; Yadava *et al.*, 1988).). The insect damage on the crop was observed on leaves, flowers, pods and seeds of all the genotypes suggesting potentially high yield losses. In a recent study conducted on pigeonpea in Nigeria, a high (78%) yield loss in pigeonpea caused by pod sucking bugs was reported (Dialoke *et al.*, 2010). Similarly, a 47% yield loss was attributed to flower thrips (Rotimi and Iloba 2008). Among the coreids (*Clavigralla* spp.), at least four different species were identified in this study (Table 3). They are considered destructive largely because the adults and nymphs feed on the developing seed while damaged mature seeds turn dark and shrivelled, rendering them unacceptable for human consumption and cannot germinate (Materu, 1970). Similarly, the *H. armigera* identified in the study can lead to significant yield losses. There are no sources of complete resistance to the pod borer that can be manipulated through breeding to develop resistant cultivars. Nonetheless, limited resistance to the pod borer complex was reported recently (Kooner and Cheema, 2010). However, it remains unclear if cultivars possessing resistance to pod borers will be palatable for humans particularly in terms of taste.

Order	Family	Species (Common Name)
1. Coleoptera	Meloidae	(i) *Mylabris oculata* (CMR beetle)
		(i) *Mylabris tincta* (Blister beetle)
2. Diptera	Agromyzidae	(i) *Melanagromyza* spp. (Podfly)
3. Heteroptera	Coreidae	(i) *Clavigralla* spp. (Tip wilter)
		(ii) *Anoplocnemis curvipes* (Black tip wilter)
		(iii) *Petalocnemis* spp. (Tip wilter)
		(iv) *Acanthocoris* spp. (Tip wilter)
	Alydidae	(i) *Riptortus* spp. (Broad-headed bug)
	Pentatomidae	(i) *Nezara viridula* (Green stinkbug)
4. Homoptera	Aphididae	(i) *Aphis craccivora* (Black cowpea aphid)
	Cicadellidae	(ii) *Empoasca fabae* (Leaf hopper)
5. Hymenoptera	Megachilidae	(i) *Megachile* spp. (Leaf-cutter bees)
6. Lepidoptera	Noctuidae	(i) *Helicoverpa armigera* (Pod borer)
7. Orthoptera	Pyrgomorphidae	(i) *Zonocerus elegans* (Elegant grasshopper)
8. Thysanoptera	Thripidae	(i) *Megalorothrips* spp. (Flower thrips)

Source: Kunjeku and Gwata, 2011

Table 3. Range of insect species occurring in the pigeonpea crop grown at Thohoyandou in the Limpopo River Basin during the 2008/09 season.

These results suggested that effective insect pest control measures would be necessary for the crop in LRB since to date, there are no improved commercial cultivars of pigeonpea that posses complete resistance to the common insect pests of economic importance. Therefore, it is necessary to establish the pest status of the various insect species in order to apply effective management strategies that utilize a wide array of natural parasites and predators such as carabids, coccinellids, anthocorids and vespids. Botanical methods (using tephrosia or neem) have been suggested but the use of broad-spectrum synthetic pyrethroids is discouraged since they can kill non-target organisms such as spiders. The effectiveness of insect pathogens such as the *Helicoverpa nuclear polyhedrosisvirus* which is regarded as relatively more friendly to the environment than chemical pesticides, is still debatable. In contrast, the use of chemical pesticides remains an effective control measure (Muthomi *et al.*, 2007).

The yield loss due to insect pests in the region can also be attributed to lack of knowledge on the part of growers in some countries. For instance, Minja *et al.*, (1999) observed that small-holder pigeonpea growers in some countries in the region did not control insect pests with conventional pesticides in the field. The majority (70%) used wood ash and about 10% used pirimiphos-methyl (Actellic dust) to protect both the grain and seed. Moreover, according to Minja *et al.*, (1999) some farmers lacked sufficient training to distinguish between the damage caused by the various insect pest groups.

Because of the relative importance of different insect species due to location, flowering time and season, future studies could focus on quantifying the damage caused by each insect type in order to determine its pest status on the crop as well as the economic threshold. In southern Africa, economic thresholds for pigeonpea pests have not been established. In other pigeonpea production regions, one larva (or three eggs) per plant can be considered as the threshold level for applying insecticides (Singh and Oswalt 1992). Likely, determining these economic thresholds would contribute to the regional optimization of the pigeonpea value chain which encompasses input supply, policy makers, farmers, harvesting, storage, processing and marketing.

4.4. Evaluation for end-use qualities

The main end-use grain qualities of pigeonpea such as grain color and size, were measured in each of the field experiments conducted at Chitedze, Ilonga, Kiboko, Ngabu and Thohoyandou as described above. The results showed that the improved cultivars possessed large (100-grain weight = 14 g), white grain (Table 1) which are preferred by end-users in the region. These types are easier to cook compared to the small-seeded types. In contrast, most of the landraces in the region possess brown (or dark) as observed for MtawaJuni even though the size may be acceptable. Nonetheless, the grain color is relatively easy to change through conventional breeding approaches. These end-use qualities are also important in the adoption of new cultivars by growers in the region (Shiferaw et al., 2007).

5. Conclusions

There is potential for improving the depressed productivity of pigeonpea smallholder cropping systems in the region by using improved cultivars. The results from the various field evaluation studies demonstrated consistently that improved cultivars can produce several fold higher yields than unimproved landraces. In addition, wilt resistant cultivars showed that even under high disease pressure, they can attain optimum grain yields across the region. While inter-cropping is popular among smallholder growers, the crop can be managed better particularly in terms of pest and weed control if it is planted as a sole crop. Ideally, the crop requires effective control of insect pests. Because of its multiple benefits, pigeonpea offers smallholder farmers in the region realistic opportunities for increasing the production of grain legumes.

Author details

E. T. Gwata[1] and H. Shimelis[2]

1 University of Venda, Department of Plant Production, Thohoyandou, Limpopo, South Africa

2 University of KwaZulu-Natal, African Center for Crop Improvement, Scottsville, South Africa

References

[1] Ae, N, Arihara, J, Okada, K, Yoshinara, T, & Johnson, C. (1990). Phosphorus uptake by pigeonpea and its role in cropping systems of the Indian subcontinent. Sci. , 248, 477-480.

[2] Agrios, G. N. (1997). Plant Pathology. Orlando, Florida, Academic Press.

[3] Amarteifio, J. O, Munthali, D. C, Karikari, S. K, & Morake, T. K. (2002). The composition of pigeonpeas [*Cajanus cajan* (L.) Millsp.] grown in Botswana. Plant Foods Hum. Nutr. , 57, 173-177.

[4] Bayaa, B, Erskine, W, & Singh, M. (1997). Screening lentil for resistance to fusarium wilt: methodology and sources of resistance. Euphytica , 98, 69-74.

[5] Beukes, D. J, Bennie, A. T. P, & Hensley, M. (1999). Optimization of soil water use in the dry crop production areas of South Africa. *In*: N. van Duivenbooden, M. Pala, C. Studer and C.L. Bielders (eds.) Proceedings of the 1998 (Niger) (April 26-30 April) and 1999 (Jordan) (May 9-13) Workshops of the Optimizing Soil Water Use (OSWU) Consortium, entitled: Efficient Soil Water Use: The Key to Sustainable Crop Production in the Dry Areas of West Asia, and North and Sub-Saharan Africa. Aleppo, Syria,ICARDA and Patancheru, India: ICRISAT. , 165-191.

[6] Dialoke, S. A, Agu, C. M, Ojiako, F. O, Onweremadu, E, Onyishi, G. O, Ozor, N, Echezona, B. C, Ofor, M. O, Ibeawuchi, I. I, Chigbundu, I. N, Ngwuta, A. A, & Ugwoke, F. O. (2010). Survey of insect pests on pigeonpea in Nigeria. J. SAT Agric. Res. , 8, 1-8.

[7] FAO ((2009). Climate and rainfall. http://wwww.fao.org/wairdocs/ilri/x5524e/ x5524e03.htmAccessed 26 February 2009).

[8]   Gwata, E. T, & Siambi, M. (2009). Genetic enhancement of pigeonpea for latitude areas in southern Africa. Afric. J. Biotechnol. , 8, 4413-4417.

[9]   Gwata, E. T, Silim, S. N, & Mgonja, M. (2006). Impact of a new source of resistance to fusarium wilt in pigeonpea. J. Phytopathol. , 154, 62-64.

[10] Gwata, E. T, & Silim, S. N. (2009). Utilization of landraces for the genetic enhancement of pigeonpea in eastern and southern Africa. J. Food Agric. Environ. , 7, 803-806.

[11] Halila, M. H, & Strange, R. N. (1997). Screening of Kabuli chickpea germplasm for resistance to fusarium wilt. Euphytica , 96, 273-279.

[12] Høgh-jensen, H, Myaka, F. A, Sakala, W. D, Kamalongo, D, Ngwira, A, Vesterager, J. M, Odgaard, R, & Adu-gyamfi, J. J. (2007). Yields and qualities of pigeonpea varieties grown under smallholder farmers' conditions in eastern and southern Africa. Afric. J. Agric. Res. , 2, 269-278.

[13] Kannaiyan, J, Nene, Y. L, Reddy, M. V, Ryan, J. G, & Raju, T. N. (1984). Prevalence of pigeonpea diseases associated with crop losses in Asia, Africa and Americas. Trop. Pest Manag. , 30, 62-71.

[14] Kooner, B. S, & Cheema, H. K. (2010). Evaluation of pigeon pea genotypes for resistance to the pod borer complex. Indian J. Crop Sci. , 1, 194-196.

[15] Kumar, R. R, Karjol, K, & Naik, G. R. (2011). Variation of sensitivity to drought stress in pigeonpea (*Cajanus cajan* (L.) Millsp.) during seed germination and early seedling growth. World J. Sci. Technol. , 1, 11-18.

[16] Kunjeku, E. C, & Gwata, E. T. (2011). A preliminary survey of insect species occurring at the stage in pigeonpea grown in Limpopo (South Africa). J. Food, Agric. Environ. , 9, 988-91.

[17] Manyasa, E. O, Silim, S. N, Githiri, S. M, & Christiansen, J. L. (2008). Diversity in Tanzanian pigeonpea [*Cajanus cajan* (L.) Millsp.] landraces and their response to environments. Gen. Res. Crop Evol. , 55, 379-387.

[18] Marston, N. L, Morgan, C. E, Thomas, G. D, & Ignoffo, C. M. (1976). Evaluation of four techniques for sampling soybean insects. J. Kansas Entom. Soc. , 49, 389-400.

[19] Materu, M. E. A. (1970). Damage caused by *Acanthomiato mentosicollis* Stal and *A. horrida* Germar (Hemiptera: Coreidae). East Afric. Agric. For. J. , 35, 429-435.

[20] Mapfumo, P, Giller, K. E, Mpepereki, S, & Mafongoya, P. L. (1999). Dinitrogen fixation by pigeonpea of different maturity types on granitic sandy soils in Zimbabwe. Symbiosis , 27, 305-318.

[21] Minja, E. M, Shanower, T. G, Songa, J. M, Ongaro, J. M, Mviha, P, Myaka, F. A, & Okurut-akol, H. (1996). Pigeonpea seed damage from insect pests in farmers fields in Kenya, Malawi, Tanzania and Uganda. Int. Chickpea Pigeonpea Newsl. , 3, 97-98.

[22] Minja, E. M. (2001). Yield losses due to field pests and integrated pest management strategies for pigeonpea- a synthesis. In: S.N. Silim, G. Mergeai, and P.M. Kimani, (eds). Proceedings of the Regional Workshop on Status and Potential of Pigeonpea in Eastern and Southern Africa 2000, Nairobi, Kenya. , 48-54.

[23] Minja, E. M, Shanower, T. G, Silim, S. N, & Singh, L. (1999). Evaluation of pigeonpea pod borer and pod fly tolerant lines at Kabete and Kiboko in Kenya. Afric. Crop Sci. J., 7, 71-79.

[24] Minja, E. M, Shanower, T. G, Songa, J. M, Ongaro, J. M, Mviha, P, Myaka, F. A, & Okurut-akol, H. (1996). Pigeonpea seed damage from insect pests in farmers fields in Kenya, Malawi, Tanzania and Uganda. Int. Chickpea PigeonpeaNewsl., 3, 97-98.

[25] Minja, E. M. (1997). Insects of pigeonpea in Kenya, Malawi, Tanzania and Uganda and grain yield losses in Kenya: a consultant's report, Bulawayo, Zimbabwe: ICRI-SAT, , 65.

[26] Mligo, J. K, & Myaka, F. A. (1994). Progress of pigeonpea research in Tanzania. In: S.N. Silim, S. Tuwafe and S. Laxman. (eds), Improvement of pigeonpea in Eastern and Southern Africa. Annual Research Planning Meeting. Bulawayo, Zimbabwe. , 51-58.

[27] Mligo, . . (2005). Adaptation and yield of pigeonpea in different environments in Tanzania. Field Crops Res. 94, 43-53.

[28] Mogashoa, K. E, & Gwata, E. T. (2009). Preliminary indicators of adaptation of pigeon pea (*Cajanus cajan* (L.) Millsp.) to Limpopo (South Africa). In: Proc. Afr. Crop Sci. Soc. 2009 Conf., Cape Town, RSA, 141 p.

[29] Muthomi, J. W, Otieno, P. E, Cheminingwa, G. N, & Nderitu, J. H. (2007). Effect of chemical spray on pests and yield of food grain legumes. Afr. Crop Sci. J. , 8, 981-986.

[30] Mzezewa, J, Misi, T, & Van Rensburg, L. D. (2010). Characterisation of rainfall at a semi-arid ecotope in the Limpopo Province (South Africa) and its implications for sustainable crop production. Water SA , 36, 19-26.

[31] Night, G, & Ogenga-latigo, M. W. (1994). Range and occurrence of pigeonpea pests in central Uganda. Afr. Crop Sci. J. , 2, 105-109.

[32] Ott, R. L. An Introduction to Statistical Methods and Data Analysis. California, Duxbury Press, (1988).

[33] Reddy, M. V, Nene, Y. L, & Kannaiyan, J. (1990). Pigeonpea lines resistant to wilt in Kenya and Malawi. Int. Pigeonpea Newsl. , 12, 25-26.

[34] Rotimi, J, & Iloba, B. N. (2008). Assessment of yield losses in tall varieties of pigeonpeasdue to the flower thrips, *Megalurothrips sjostedti* Trybom. (Thysanoptera: Thripidae) Biosci. Res. Comm. , 20, 305-308.

[35] SAS InstituteSAS user's Guide: Statistics. Cary, NC, SAS Institute, (1989).

[36] Shanower, T. G, Romeis, J, & Minja, E. M. (1999). Insect pests of pigeonpea and their management. Annual Rev. Entomol. , 44, 77-96.

[37] Silim, S. N, Coe, R, Omanga, P. A, & Gwata, E. T. (2006). The response of pigeonpea genotypes of different duration types to variation in temperature and photoperiod under field conditions in Kenya. J. Food. Agric. Envir. , 4, 209-214.

[38] Silim, S. N, Bramel, P. J, Akonaay, H. B, Mligo, J. K, & Christiansen, J. L. (2005). Cropping systems, uses, and primary *in situ* characterization of Tanzanian pigeonpea (*Cajanus cajan* (L.) Millsp.) landraces. Genet. Res. Crop Evol. , 52, 645-654.

[39] Singh, F, & Oswalt, D. L. (1992). Insect pests of pigeonpea. In: Pest management Skills Development Series ICRISAT, Patancheru, India. (12), 6-20.

[40] Songok, S, Ferguson, M, Muigai, A. W, & Silim, S. (2010). Genetic diversity in pigeonpea [*Cajanus cajan* (L.) Millsp.] landraces as revealed by simple sequence repeat markers. Afric. J. Biotechnol. , 9, 3231-3241.

[41] Spurgeon, D. W, & Cooper, W R. (2011). Among-sampler variation in sweep net samples of adult *Lygushesperus* (Hemiptera: Miridae) in cotton. J. Econ. Entomol. , 104, 685-692.

[42] Tekeoglu, M, Tullu, A, Kaiser, W. J, & Muehlbauer, F. J. (2000). Inheritance and linkage of two genes that confer resistance to fusarium wilt in chickpea. Crop Sci. , 40, 1247-1251.

[43] Whiteman, E. S, Byth, D. E, & Wallis, E. S. (1985). Pigeonpea *In*: R.J. Summerfield, R.H. Ellis (eds). Grain Legume Crops. Collins, London. , 658-698.

[44] Yadava, C. P, Lal, S. S, & Sachan, J. N. (1988). Assessment of incidence and crop losses due to pod-borers of pigeonpea (*Cajanus cajan*) of different maturity groups. Indian J. Agric. Sci. , 53, 216-218.

Chapter 5

Evaluation of the Agronomic Performance of Vernonia (*Vernonia galamensis*) Germplasm

H. Shimelis and E.T. Gwata

Additional information is available at the end of the chapter

1. Introduction

Vernonia [*Vernonia galamensis* (Cass.) Less.; 2n = 18] is a relatively new crop in many parts of Africa. Most of *Vernonia* species occur in South America but more than 300 species from Africa have been described with most occurring in Ethiopia and Madagascar. Apart from these two countries, vernonia is also grown in Cape Verde, Eritrea, Mozambique, northern Tanzania and Senegal (Fig. 1). The greatest diversity of vernonia is found in east Africa while a single variety occurs in West Africa. The genus vernonia comprises of more than a thousand species which vary from annual herbs and shrubs to perennial trees (Baye *et al.*, 2001). There are six major subspecies namely *afromontana, galamensis, gibbosa, nairobensis, lushotoensis* and *mutomonesis*. Among these, *galamensis* shows the highest genetic diversity (Gilbert, 1986). It contains four botanical varieties namely *australis, ethiopica, galamensis* and *petitiana* (Gilbert, 1986).

The morphological characters of the Vernonia plant were described comprehensively by Perdue et al., (1986). The authors described Vernonia as herbaceous, usually annual, varying from small ephemerals 20 cm tall with a single flower head to robust rather diffusely branching somewhat shrubby plants which grow up to 5 m tall with many flower heads. The authors also noted that the stems branch only after the first flower head is formed and the inflorescence consists of a terminal flower head with lateral flower heads from the uppermost axils. The leaves alternate and are membraneous, 0.6-5.0 cm wide, up to 25 cm long (Perdue *et al.*, 1986). The classification of the species into six subspecies is based on characters of the phyllaries (Perdue *et al.*, 1986).

Vernonia could potentially grow as a seed oil crop in tropical and subtropical environments with frost-free and short-day length for flower initiation and development. For instance, the crop was grown successfully in Zimbabwe where seed yields varied from 1.7 to 2.5 t/ha during 1986 to 1987. Thus far there are no released cultivars of *V. galamensis*. Development of improved varieties and production technologies are still in the early stages. The present study is based on germplasm collections from Ethiopia, the center of diversity for *V. galamensis* var. *ethiopica*. The study identified agronomically promising genetic resources useful in further development of more productive cultivars. Further agronomic and utilization research on *V. galamensis* needs to continue before it can be fully-established as a new crop.

In the US Vernonia domestication and large-scale production an oilseed crop was limited due to short-day length requirement for flower initiation and development. In these environments, frosts following flowering inhibits complete seed development and maturity. The crop also performs poorly in areas with excessive moisture, poor soil drainage and insufficient length of growing season. Dierig and Thompson (1993) indicated several barriers that limit full domestication, cultivation and production in the United States such as day-neutrality, autofertility, non-dormant seed germination, good seed retention, increased uniformity of seed maturity, and high oil and vernolic acid contents (Dierig and Thompson 1993).

The successful production of the crop requires well-drained and porous soils. In contrast, in poorly-drained soils, terminal growth is severely retarded and can stop before flowering. The upper portion of the plant dies and branches subsequently grow from the base of the plant, but also wither and die without flowering. Soil with intermediate drainage will produce plants that develop a few flower heads, but with low seed yields. Vernonia seed is planted directly into the soil at a depth of 1 to 2 cm and spaced at 60 cm (intra-row) x 60 cm (inter-row). A firm, level and weed free seedbed is necessary for enhancing rapid establishment and good stand. The seed is relatively small and often with poor germination. Therefore deeper planting is discouraged.

Weed control is essential in the early field establishment due to the poor seedling vigor of Vernonia. The weeds are manually controlled and recommendations on chemical weed controls are not available. No herbicides are currently registered for use in Vernonia. In our studies Vernonia has no major serious diseases and insect pests threatening the crop. Vernionia heads should be harvested when the plants show complete leaf senescence. The seed stays on the heads of the plants for 30 to 45 days after ripening. Immediately after harvesting, the heads are threshed manually in order to obtain the seed.

Although vernonia is cultivated in many tropical countries, the full potential of the crop as an oilseed is yet to be exploited in Africa and elsewhere. It is potentially a useful industrial oil seed crop for the production of natural epoxy oil (Thompson *et al.*, 1994a; Mohamed *et al.*, 1999). The seeds of vernonia produce naturally exposidized oil consisting of vernolic acid, palmitic acid, linoleic acid, arachidic acid, linolenic acid and steraic acid (Carlson *et al.*, 1981; Ayorinde *et al.*, 1988). Vernolic acid is the dominant fatty acid. Because of the production of the naturally exposidized oil, the industrialized processing of this oil is inexpensive. It is also friendly to the environment largely because it does not emit volatile organic compounds. In addition, the oil can be stored at sub-zero temperatures.

Vernonia oil is used in a variety of ways in the chemical industry. The seed of this crop produces useful natural epoxy fatty acids that are better than artificial epoxy oils. In addition, the vernonia oil from the seed contains a wide range of fatty acids such as vernolic acid, linoleic acid, oleic acid, palmitic acid and stearic acid which have industrial uses. The oil is useful in the manufacture of polyvinylchloride and structural polymers for the production of plastic materials (Mebrahtu *et al.*, 2009) and petrochemicals. The cake formed after oil extraction is high (43.75%) in crude protein and is suitable for animal feed.

Epoxy oils have wide industrial applications, such as in plasticizers, additives in flexible polyvinyl chloride, synthesis of epoxy resins, adhesives, and insecticides. The triglyceride oil rich in vernolic acid, is environmentally friendly, less expensive and less viscous compared to other artificial epoxy oils (Thompson *et al.*, 1994b; Mohamed *et al.*, 1999). Vernolic acid makes up 72 to 80% of the acids present in the seed oil. Vernonia oil also contains other fatty acids, such as linoleic acid (12–14%), oleic acid (4–6%), stearic acid (2–3%), palmitic acid (2–3%), and a trace amount of arachidic acid (Carlson *et al.*, 1981; Ayorinde *et al.*, 1988).

In spite of its multiple uses, the cultivation and commercialization of vernonia is limited by several factors. Firstly, the seed of vernonia does not attain maturity uniformly. Secondly, the pods shatter easily at maturity leading to significant yield loss. Thirdly, the vernonia plant is generally tall. To date, the crop has not been adapted to mechanized harvesting, seed threshing and cleaning. The harvesting and seed processing are carried out manually.

Due to the high oil and vernolic acid content and its relatively low shattering nature, subsp. *galamensis* var. *ethiopica* has been the major focal point of research aimed at domestication and commercialization of the crop. Viable production of vernonia as an alternative industrial crop in marginal tropical and subtropical areas by smallholder or commercial farmers hinges on the identification of suitable varieties that are adapted to the prevalent cropping systems. Alternatively, improved cultivars that are high yielding could also be adopted by farmers. However, there is a dearth of information on the agronomic performance of the current germplasm of vernonia. There are no improved cultivars adapted to the dryland tropical areas in southern Africa such as the Limpopo Province (South Africa). In such areas, vernonia could provide a source of raw materials for agro-processing industries. In addition, vernonia cultivation could provide a significant diversification of the existing cropping systems in the region. In the Limpopo Province of South Africa for instance, the climate is semi-arid and characterized by low mean annual rainfall (300 to 600 mm) with a predominantly sandy-loam soil with reduced fertility (Thomas, 2003). The rainfall pattern is highly variable in some ecotopes in the area (Mzezewa *et al.*, 2010). These harsh agro-ecological conditions could be suitable for the domestication of vernonia and detailed investigations in the agronomic performance of the species would be necessary before large-scale production of the crop in the area is recommended to growers. Therefore objectives of this Chapter are three-fold. The first section focuses on the evaluation of the agronomic performance of *V. galamensis* var. *ethiopica* that was conducted in the in Limpopo Province of South Africa. The second section focuses on the selection of germplasm for high-quality and high-quantity oil that may be used further in strategic breeding of the crop aimed at developing an alternative industrial oil crop in the region and similar environments. The third segment of the Chapter examines the implications and recommendations for cropping systems particularly in the dry land areas in southern Africa.

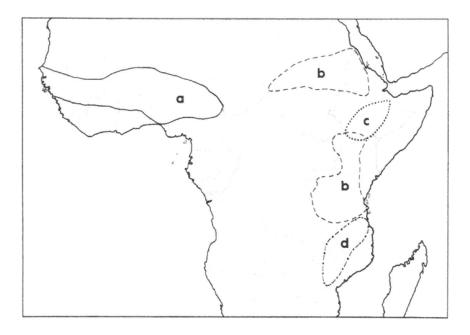

Figure 1. The general distribution of *V. galamensis*. a = West Africa region (Extending from Guinea through Ivory Coast, Mali, Bukino Faso, Ghana to Nigeria) b = Sudan, Kenya and Tanzania region; c = central-southern Ethiopia; d = 4 Malawi, Mozambique and eastern Zimbababwe region). (Source: Adapted from Perdue et al., 1986).

2. Agronomic performance of *Vernonia* germplasm

In this field evaluation, 36 accessions of *Vernonia galamensis* subsp. *galamensis* var. *ethiopica* (Table 1) were obtained from the Biodiversity Institute (Ethiopia). The eastern and south-eastern region of Ethiopia is considered as a natural habitat for *V. galamensis* subsp. *galamensis* var. *ethiopica* (Gilbert, 1986) in which this botanical variety is believed to be most diverse. The habitat is located at an altitude of 700 to 2400 m and receives little seasonal rainfall (about 200 mm). The soils are poor (infertile). Prior to the study, the homozygosity of each accession was maintained through three cycles of selections (Shimelis *et al.*, 2008). In this study, the field experiments were established in Limpopo Province of South Africa under rain fed conditions during the 2006, 2007 and 2008 growing seasons. Experiments were established at two localities namely at Syferkuil (23° 84' S and 29° 71' E) and Gabaza. Syferkuil is situated at an altitude of 1261.6 m above sea level (asl). It has annual maximum temperature ranging from 28-30 °C and receives an average annual rainfall of 468 mm. This site has sandy loam soil, of the Hutton form, Glenrosa family, with the pH ranging from 6 - 6.2. While Gabaza (23° 50' S 30° 10' E) has an altitude of 1100 m asl with an annual average rainfall of 600 mm. At Gabaza the annual temperature ranges from 15 - 37 ℃ and with clay-loam soil. In general, soil, climatic, and biological conditions of the two locations varies considerably. However, both locations are not prone to frost and hence suitable representative agro-ecologies conducive for Vernonia production in the Limpopo region.

Each experiment was laid out as a partially balanced lattice design with six incomplete blocks replicated three times. Each block was (21.6 m²) 3 m wide x 7.2 m long and consisted of two rows spaced at 0.6m apart. Within the row, the seeds were planted at 0.6 m from each other. During planting and flower head initiation, fertilizer (12.5% N, 8.3% P, 4.2% K and 0.5% Zn) was split applied manually at a rate of 30 kg ha⁻¹ (Shimelis *et al.*, 2008). In literature, there are no reports on fertilizer response and nutrient requirement in *V. galamensis*. Preliminary observations in Ethiopia showed that the application of 150 kg/ha nitrogen enhances lodging. Thus only a maintenance amount of 20 kg/ha for N and P each was supplied (Baye et al. 2001).

2.1. Agronomic measurements

The agronomic traits that were measured included the duration (in days) to 50% flowering (50%DF), the number of productive primary heads (PPH) per plant, the number of productive secondary heads (PSH) per plant, the thousand seed weight (1000-SW) and seed yield (SY). Measurements for each accession were done on 10 plants that were selected randomly and tagged in each block within a replication. In both seasons (2005 and 2006) the field evaluation was conducted during the summer (January to June) cropping season.

Oil content (OC) (based on dry seed weight) was measured. Total lipid (TL) was extracted from ground seeds following the method of Folch *et al.* (1957), with a chloroform to methanol ratio of 2:1. An antioxidant, butylated hydroxytoluene, was added at a concentration of 0.001% to the chloroform–methanol mixture. Total extractable fat was determined gravimetrically and expressed as percent fat (%F) (w/w) per 100-g sample. The fatty acid composition was determined after transesterification of the extracted lipid by the addition of tri-methyl sulphonium hydroxide (Butte, 1983) and quantified using the gas chromatograph technique (Shimelis *et al.*, 2008). Because of insignificant variations between the two seasons, the average of the data sets over the two seasons was used for statistical analysis. The data sets for each quantitative character were analysed using the SAS GLM procedure for a fixed model with the SAS software version 9.1.3 (SAS 2004) followed by mean separation using Fisher's protected LSD. Phenotypic correlations between agronomic traits, oil content and fatty acids were determined using simple Pearson correlation.

The results of this evaluation showed significant differences among *V. galamenisis* var. *ethiopica* accessions for agronomical traits, oil and fatty acids content. The crop thrived at the location indicating its adaptation to the prevailing arid conditions (Fig. 2). Accessions Vge-10 and Vge-19 were relatively early flowering requiring 88 d to attain 50% flowering. In contrast, the accessions Vge-15 and Vge-36 flowered late (>140 d) (data not shown). In terms of plant height, four accessions namely Vge-18, Vge-19, Vge-25 and Vge-30 were relatively short, averaging about 133.6 cm. This indicated that the Vge-19, which was collected from eastern region of Ethiopia (Table 1), could be selected for both earliness and reduced height. In similar studies involving *V. galamenisis*, Angelini *et al.* (1997) and Bhardwaj *et al.*, (2000) reported variations

Accession Code	Place of Origin	Location
Vge-1		
Vge-2	Bedeno	09°06' N, 41°38' E
Vge-3		
Vge-4		
Vge-5		
Vge-6	Melkabelo	09°12' N, 41°25' E
Vge-7		
Vge-8		
Vge-9		
Vge-10	Harar Zuria	09°19' N, 42°07' E
Vge-11		
Vge-12		
Vge-13		
Vge-14	Metta	09°25' N, 41°34' E
Vge-15		
Vge-16		
Vge-17	Gelemso	08°49' N, 40°31' E
Vge-18		
Vge-19		
Vge-20		
Vge-21	Yirgalem	06°42' N, 038°21' E
Vge-22		
Vge-23		
Vge-24	Leku	06°52' N, 038°27' E
Vge-25		
Vge-26		
Vge-27		
Vge-28	Awassa	06°52' N, 038°27' E
Vge-29		
Vge-30		
Vge-31	Areka (06°48' N, 037°43' E)	Areka (06°48' N, 037°43' E)
Vge-32		
Vge-33		
Vge-34		
Vge-35	Arsi-Negele	(07°00' N, 038°35' E)
Vge-36		

Source: Adapted from Shimelis *et al.,* 2008

Table 1. Vernonia germplasm evaluated at the University of Limpopo Experimental Farm (Syferkuil).

in duration to flowering and plant height. In comparison with the other accessions, Vge-19 obtained a high number of PPH (72). On the other hand, accession Vge-11 showed a relatively high number of PSH (35) per plant (Table 2).

Accession Code	Agronomic Trait					
	50%DF (d)	PHT (cm)	PPH	PSH	1000-SW (g)	SY (kg/ha)
Vge-17	93.33	143.00	58	27	1.68	3126.09
Vge-18	98.33	131.60	53	25	2.07	3016.86
Vge-19	87.67	133.26	72	30	1.98	2871.00
Vge-16	93.33	151.00	60	31	1.84	2904.34
Vge-12	98.33	150.27	53	21	2.31	2706.00
Vge-4	98.33	164.73	35	16	2.58	2695.44
Vge-30	100.00	135.00	55	14	3.54	2658.48
Vge-25	111.00	134.73	63	14	3.27	2647.26
Vge-11	99.33	151.80	62	35	2.49	2640.00
Vge-27	116.33	156.00	56	16	2.62	2587.20
Mean	99.60	145.14	57	23	2.44	2785.27

Table 2. Agronomic performance of the best 10 (based on seed yield) vernonia accessions that were evaluated for six traits during 2005 and 2006 in Limpopo Province (South Africa). (50%DF = number of days to 50% flowering; PHT = plant height; PPH = number of primary productive heads; PSH = number of secondary productive heads; 1000-SW = one thousand seed weight; SY = seed yield).

Under these agro-ecological conditions at the testing location, some of the germplasm was more productive in terms of the number of mature seed heads than reported by Bhardwaj *et al*,. (2000) who found twice as many immature seed heads (60 – 80) compared to mature seed heads (20 - 43) per plant. The seed yield was relatively high in accession, Vge-17, Vge-18, and Vge-19 averaging about 3.0 t ha^{-1} (Table 2). This was consistent with findings from other researchers working with germplasm of vernonia originating from east Africa (Thompson *et al*., 1994a; Mohamed *et al*., 1999; Baye *et al*., 2001). The duration to flowering as measured by 50%DF showed a poor association with 1000–SW but the number of PPH showed significant positive correlations with both seed yield and the number of productive secondary seed heads suggesting that improved seed yield in this species could be achieved through simultaneous selection of increased number of productive heads.

The oil analysis indicated considerable variation in the seed oil content with accession Vge-4 attaining the (35.86%) (Table 3). The observed variation in oil content in this study concurred with observations from other similar studies (Mohamed *et al*., 1999; Angelini *et al*., 1997). The fatty acid profiles showed vernolic acid (VA) content ranging from 72.21 to 77.06% (Table 3). Four accessions from the eastern region of Ethiopia (Vge-6, Vge-8, Vge-9, and Vge-11) and two

Figure 2. *Vernonia galamensis* var. *ethiopica* thriving at a semi-arid location at Syferkuil (Limpopo Province, South Africa).

from the south (Vge-25 and Vge-35) obtained superior yield of vernolic acid. In addition, there was considerable variation in the proportion of individual fatty acids among the accessions. For instance, the stearic acid in accession Vge-33 was about 73% of the palmitic acid but was almost equal (95%) to quantity of stearic acid in the accession (Fig. 3).

In similar studies, VA content was low (Angelini *et al.*, 1997; Mohamed *et al.*, 1999; Bhardwaj *et al.*, 2000). Linoleic acid ranged from 12.05 to 14.73%. The highest oil yield (966.58 kg ha^{-1}) was observed for accession Vge-4 which jointly with accession Vge-18 produced good seed yield. A significant positive correlation between 1000-SW and seed oil content was found indicating that the accessions with relatively heavier seed contained increased levels of oil content. However, there were poor associations between oil yield and other traits.

While the information regarding the heritability of these traits in vernonia is fragmentary at best, the variation and association between agronomic traits with seed oil and fatty acids suggested that there is merit in exerting effort aimed at the genetic improvement of these traits in vernonia. This could help in diversifying the existing African cropping systems since vernonia can be used as a cash crop in the production of natural epoxy oils. Industries produce epoxy oils by modification of petrochemicals and epoxidation of oils from seeds of soybean [*Glycine max* (L.) Merr.] and linseed (*Linum ustitatissimum* L.). However, artificially epoxidized oil is expensive and contains volatile organic solvents with high emission to the environment during processing and use. The release of volatile organic solvents into the air causes environmental pollution. In addition, the oil is highly viscous and cannot be stored below 0°C. The natural oil from vernonia has unique properties due to its non-volatility and low viscosity.

Accession Code	Trait			
	SO	VA	LA	OA
Vge-4	35.86	75.37	13.52	4.55
Vge-3	34.19	72.99	13.19	5.09
Vge-32	33.11	75.88	13.56	3.99
Vge-33	30.89	75.19	14.09	4.18
Vge-30	30.85	75.97	13.66	3.86
Vge-21	29.46	74.29	13.19	5.40
Vge-6	29.40	77.06	12.58	4.16
Vge-31	29.29	75.70	14.36	4.51
Vge-22	29.18	72.21	12.81	5.16
Vge-34	29.16	76.17	12.05	3.78
Mean	31.14	75.08	13.30	4.47

Table 3. Mean response of oil content and fatty acid composition (%) among the best 10 (based on seed oil) of vernonia that were evaluated during 2005 and 2006 in Limpopo Province (South Africa). (SO = seed oil; VA = Vernolic acid; VA = Linoleic acid; OA = oleic acid).

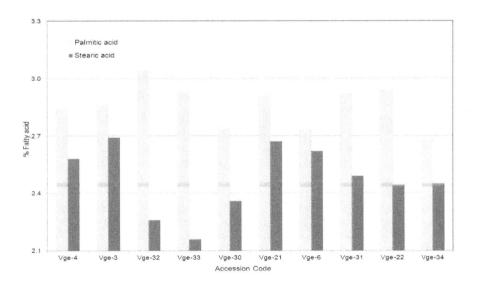

Figure 3. Palmitic and stearic acid levels among selected 10 accessions of vernonia that were evaluated during 2005 and 2006 in Limpopo Province (South Africa).

According to Ray (1994), when used as a solvent in alkyd-resin paint, the natural oil becomes part of the dry paint surface due to its reactive dilutant properties, preventing emission of volatile organic solvents. Vernolic acid from plants was first discovered and isolated from *V. anthelmintica* (L.) Wild., native to India (Gunstone, 1954). This species had excessive seed shattering, which prevented its further development. The collection and subsequent evaluation of V. *galamensis* was found promising because of the quantity and quality of the seed oil and seed retention (Carlson *et al.*, 1981; Thompson *et al.*, 1994b; Perdue, 1988). The germplasm identified in this evaluation is currently being used in the strategic improvement of vernonia to maximize seed yield and oil content as an alternative crop in the province and similar environments.

3. Environmental influence on agronomic performance

In its natural habitat, vernonia thrives as a weed under marginal low rainfall and poor soil fertility (Gilbert, 1986). There are no major pests and diseases that limit the production of this crop. Therefore, vernonia is an alternative industrial crop particularly in low input tropical cropping systems such as prevalent in the semi-arid environments in southern Africa. Typically, the marginal areas are inhabited by smallholder growers who have limited resources for crop production. Crop failure due to moisture deficits in these areas is common. However, diversification of the cropping systems often reduces the risks of crop failure.

This component of the study examined the genotype x environment (Helgadóttir and Kristjánsdóttir, 1991; Lin and B inns, 1988) interaction of seed and oil yield in vernonia using 10 selected lines. Field experiments were established as described above (see Section 2). At maturity, both primary and secondary heads were harvested per plot. The seed yield was measured in three replicates over three years and at two locations. The oil content, total lipids and extractable fat content were determined and analysed as described above.

The results showed significant interaction (P≤0.05) between genotype x location for seed yield, seed oil content and oil yield (Table 4). Differential responses of the genotypes for the traits were detected across locations or years. At Gabaza, genotype Vge-18 obtained the highest seed yield (3337 kg/ha) during the 2008 cropping season (Table 5). Similarly, Vge-18 and Vge-17 attained relatively high seed yield at Syferkuil. The presence of genotypic variability and genotype x environment interaction for seed yield in vernonia was reported in other studies (Thompson *et al.*, 1994a; Mohammed *et al.*, 1999; Baye *et al.*, 2001). These authors found variations in seed yield among *Vernonia galamensis* collections from East Africa that ranged from 60 to 2800 kg/ha. In this study, accession Vge-17 also showed stable performance across locations and years (Table 5). At Gabaza, accession Vge-4 obtained higher yield than at Syferkuil (Table 6). The relatively high oil yield in Vge-18 was attributed partly to the high seed yield. The study found significant variations in seed yield (1990–3337 kg/ha), oil content (25–43%) and oil yield (675–1370 kg/ha) among *V. galamensis* var. *ethiopica* selections when tested across two the locations over a three year period under the agro-ecological conditions in Limpopo province (South Africa).

Source of Variation	df	Mean Square		
		Seed Yield	Oil Content	Oil Yield
Genotypes (G)	9	115901.58**	255.24**	94298.67**
Locations (L)	1	167944.80**	4450.54**	3988343.47**
Years (Y)	2	203707.89**	17.43**	62339.18**
G x L	9	52895.19**	75.33**	89664.68**
G x Y	18	25766.95*	3.25ns	8245.82*
L x Y	2	36358.72ns	1.70ns	6375.82ns
G x L x Y	18	7332.29ns	2.82ns	3382.90ns
Replication within L and Y	12	24812.48	2.22	4996.08
Error	108	13304.50	3.18	3753.20
Total	179			

* = significant at the 5% probability level

** = significant at the 1% probability level

ns = not significant at the 5% probability level

Table 4. Mean squares for seed yield, of oil and oil yield among 10 selected Vernonia accessions evaluated over three cropping seasons in in Limpopo Province (South Africa).

Accession Code	Seed Yield (kg/ha)											
	Gabaza						Syferkuil					
	2006	Rank	2007	Rank	2008	Rank	2006	Rank	2007	Rank	2008	Rank
Vge-3	2275.07	10	2300.11	10	2208.75	10	2072.87	10	1993.42	10	1989.67	10
Vge-4	2409.91	8	2518.33	7	2552.83	8	2620.49	6	2649.83	6	2517.41	7
Vge-12	2918.67	4	2761.67	5	2903.33	4	2732.00	4	2688.79	4	2860.58	4
Vge-16	2922.00	3	2880.33	3	3064.33	3	2783.00	5	2674.00	5	2867.08	3
Vge-17	3085.00	2	3006.33	2	3137.33	2	3217.03	1	2914.74	1	3179.42	1
Vge-18	3118.75	1	3095.33	1	3337.33	1	2990.29	2	2818.75	2	3152.29	2
Vge-19	2791.75	5	2774.00	4	2899.60	5	2872.00	3	2718.00	3	2806.00	5
Vge-30	2623.74	6	2623.74	6	2623.74	6	2623.83	7	2450.00	7	2621.49	6
Vge-32	2385.33	9	2477.33	8	2503.00	9	2345.28	8	2420.67	8	2503.67	8
Vge-33	2442.41	7	2414.00	9	2525.75	7	2375.08	9	2203.33	9	2488.89	9
Mean	2697.26		2685.12		2772.60		2663.19		2553.15		2698.65	

Table 5. Mean seed yield among 10 selected vernonia accessions that were evaluated over three cropping seasons at Gabaza and Syferkuil in Limpopo Province (South Africa).

Accession Code	Oil Content											
	Gabaza						Syferkuil					
	2006	Rank	2007	Rank	2008	Rank	2006	Rank	2007	Rank	2008	Rank
Vge-3	41.95	4	40.11	5	40.47	9	34.33	2	34.69	2	35.27	1
Vge-4	42.55	1	42.28	1	46.62	3	34.78	1	34.81	1	35.14	2
Vge-12	42.16	3	41.25	4	41.80	6	28.12	6	27.38	6	27.27	7
Vge-16	39.83	7	39.04	9	42.02	4	26.33	7	27.15	6	29.30	6
Vge-17	26.37	10	26.33	10	27.56	10	24.69	10	24.60	7	24.61	10
Vge-18	35.82	9	39.94	8	41.04	8	25.88	8	25.70	10	25.27	9
Vge-19	41.42	5	39.96	7	41.84	5	25.57	9	25.50	8	26.75	8
Vge-30	41.32	6	41.32	3	41.32	7	31.17	4	30.65	9	32.00	5
Vge-32	38.99	8	42.00	2	42.77	2	33.22	3	34.09	4	33.22	3
Vge-33	42.17	2	40.25	6	43.22	1	30.50	5	30.65	3	32.52	4
Mean	39.26		39.23		40.36		29.46		29.52	5	30.14	

Table 6. Mean oil content among 10 selected vernonia accessions that were evaluated over three cropping seasons at Gabaza and Syferkuil in Limpopo Province (South Africa).

In a separate study conducted in the greenhouse, Shimelis *et al.*, (2006) reported similar variation in oil content variation (24–29%), vernolic acid (73-77%), linoleic acid (12-14%), oleic acid (3.5-5.5%), palmitic acid (2.4-2.9%) and stearic acid (2.3-2.8%). These findings demonstrated the genetic potential of vernonia as an alternative industrial oil crop in the region. The crop could eventually supersede petrochemicals and oils that are artificially epoxidized and emit volatile organic solvents which pollute the environmental.

4. Cultivar selection criteria

Further analyses of the genotypic correlations and path analysis (Wright, 1934; Li, 1956; Dewey and Lu, 1959; Bhatt, 1973; Kang *et al.*, 1983) in order to determine relationships between seed yield and seed oil content with other important agronomic traits among the 36 diverse accessions of vernonia (*V. galamensis* var. *ethiopica*) were carried out (Shimelis and Hugo, 2011). The information derived from such analyses is useful in identifying the best indirect selection criteria and optimizing the efficiency of selecting particularly for increased seed yield and oil content in vernonia.

A relatively high direct path coefficient value (0.49) and a highly significant genotypic correlation ($r_g = 0.81$, $P < 0.01$) were exhibited between seed yield and the number of PPH. This association from the direct path value indicates that PPH can be utilized as the first principal selection criterion for improving seed yield among these accessions. In addition, the analysis indicated that selecting for increased number of primary heads would tend to reduce simultaneously the duration to maturity as well as plant height in contrast with selecting directly for 50%DF and plant height. The analysis also showed that improvement in oil content in vernonia can be obtained through the simultaneous selection of lines that display high seed yield and seed weight. If only genotypic correlations were considered, early flowering, short plant stature and high number of productive heads could be useful selection criteria for achieving high seed yield in *V. galamensis* var. *ethiopica*.

There is no adequate information regarding associational studies on the agronomic traits in vernonia. Bhardwaj et al. (2000) reported non-significant correlations among seed yield and oil content while Baye and Becker (2005) reported a positive correlation between seed yield and seed weight in vernonia. In summary, selection for increased number of PPH is recommended as the principal selection criterion for improving seed yield in this species. The selection for 1000-SW and increased seed yield can be regarded as major selection criteria for enhancing oil content in *V. galemanesis*.

5. Prospects for genetic improvement of Vernonia

The main findings from the research work done so far suggest that some of the production and commercialization of the species can be addressed through a combination of conventional and molecular breeding approaches. For instance, plant height can be reduced routinely using standard modern plant breeding approaches. Similarly, the dehiscence of the heads can also be improved through breeding. This has been demonstrated in several other field crops (Kadkol et al., 1989; Morgan et al., 2000). Moreover, source of shattering resistance were identified in *V.* subsp. *galamensis* var. *ethiopica*. According to Morgan *et al.*, (1988), in rapeseed, increased shatter resistance is desirable partly because it can delay harvesting in order to allow more even maturing of seeds and decrease the incidence of chlorophyll contamination from immature seeds in extracted oil. The variability in the traits of economic interest such as seed yield and oil yield of vernonia indicated that there is potential to select for increased levels of these traits. In other species, oil yield was manipulated through plant breeding approaches (Cahoon. 2003; Cahoon et al., 2007; Wittkop et al., 2009).

The various modern molecular tools that are applied in plant breeding (Li *et al.*, 2010; Paux *et al.*, 2010; Ramalema *et al.*, 2010; Raman *et al.*, 2010; Zhao *et al*; 2007; Slade *et al.*, 2005; Sharma *et al.*, 2001; Tanksley *et al.*, 1989) also offer exciting prospects for the genetic improvement of vernonia. While the development and use of transgenic field crops for food remains controversial in many parts of the world partly because of the perceived risks on human and animal health, transgenic cultivars of vernonia that are high yielding in industrial grade oils should probably find some support even among the critics of the technology. Shimelis *et al.*, (2006) showed the feasibility of raising vernonia in a sheltered environment such as a greenhouse where, to all intents and purposes, undesirable gene flow (for instance from transgenic vernonia) can be prevented. High value crops, particularly horticultural and ornamental species, are routinely produced in controlled environments in many parts of the world.

The demand in the petro-chemical industries for the high quality of oils from this crop is likely to attract capital investment into the marketing and commercial production of vernonia in the marginal areas in Africa. Value addition industries for vernonia such as those for producing bio-based chemicals (Hatti-Kaul *et al.*, 2007) in the region, would be useful for economic development.

In conclusion, we believe that there is merit in investing in the genetic enhancement of vernonia in Africa since the oils from the seed of this crop have numerous advantages over comparative by-products from other crops. There is ample evidence that the crop is adapted to the harsh agro-ecological conditions prevalent in Limpopo Province and beyond. Likely, improved cultivars that are non-shattering, high yielding and mature uniformly will be adopted more widely by growers in the region.

Acknowledgements

The authors would like to express gratitude to their respective Institutions and the National Research Foundation (South Africa) for the financial support that was used to fund this work. The assistance with graphics and typing rendered by Ms Muno Gwata is gratefully acknowledged.

Author details

H. Shimelis[1] and E.T. Gwata[2]

1 University of KwaZulu-Natal, African Center for Crop Improvement, Scottsville, South Africa

2 University of Venda, Department of Plant Production, Limpopo, South Africa

References

[1] Agrobase. (2005). Agrobase generation II user's manual. Agronomix Software, Manitoba, Canada.

[2] Angelini, L.G., Moscheni, E., Colonna, G., Belloni, P., and Bonari, E. (1997). Variation in agronomic characteristics and seed oil composition of new oilseed crops in central Italy. Ind. Crops Prod. 6,313–323.

[3] Ayorinde, F.O., Osman, J.G., Shepard, R.L. and Powers, F.T. (1988). Synthesis of azelaic acid and suberic acid from *Vernonia galamensis* oil. J. Am. Oil Chem. Soc. 65,1774–1776.

[4]   Baye, T. and Becker, H. (2005). Genetic variability and interrelationship of traits in the industrial oil crop *Vernonia galamensis*. Euphytica 142, 119–129.

[5]   Baye, T., Kebede, H. and Belete, K. (2001). Agronomic evaluation of *Vernonia galamensis* germplasm collected from eastern Ethiopia. Ind. Crops Prod. 14,179–190.

[6]   Bhardwaj, H.L., Hamama, A.A., Rangappa, A. and Dierig, D.A. (2000). *Vernonia* oilseed production in the mid-Atlantic region of the United States. Ind. Crops Prod. 12,119–124.

[7]   Bhatt, G.M. (1973). Significance of path coefficient analysis in determining the nature of character association. Euphytica 22,338–343.

[8]   Butte, W. (1983). Rapid method for the determination of fatty acid profiles from fats and oils using trimethyl sulphonium hydroxide for transesterifi cation. J. Chromatogr. 261,142–145.

[9]   Carlson, K.D., Schneider, W.J., Chang, S.P. and Princen, H. (1981). *Vernonia galamensis* seed oil: A new source for epoxy coatings. *In*: Pryde, E.H., Princen, L.H. and Mukherjee, K.D. (eds). New Sources of Fats and Oils. Am. Oil Chem. Soc., Champaign, IL., pp. 297–318.

[10]   Cahoon, E.B. (2003) Genetic enhancement of soybean oil for industrial uses: prospects and challenges. AgBioForum 6,11–13.

[11]   Cahoon, E.B., Shockley, J.M., Dietrich, C.R., Gidda, S.K., Mullen, R.T. and Dyer, J.M. (2007) Engineering oilseeds for sustainable production of industrial and nutritional feedstocks: solving bottlenecks in fatty acid flux. Curr. Opin .Biotechnol. 10, 236–244.

[12]   Dewey, D.R. and Lu, K.H. (1959). Acorrelation and path coefficient analysis of components of crested wheat grass seed production. Agron. J. 51,515–518.

[13]   Dierig, D.A., Coffelt, T.A., Nakayama, F.S. and Thompson, A.E. (1996). Lesquerella and vernonia: oilseeds for arid lands. *In*: J. Janick (ed). Progress in New Crops. ASHS Press, Alexandria, VA. pp. 347–354.

[14]   Folch, J., Lees, M. and Sloane-Stanley, G.H. (1957). A simple method for the isolation and purification of total lipids fromanimal tissue. J. Biol. Chem. 226,497–509.

[15]   Gilbert, M.G. (1986). Notes on East African Vernonieae (Compositae). A revision on the *Vernonia galamensis* complex. Kew Bull. 41,19–35.

[16]   Gunstone, F.D. (1954). Fatty acids: Part II. The nature of the oxygenatedacids present in *Vernonia* anthelmintica (Willd.) seed oil. J. Chem. Soc. (May), 1611–1616.

[17]   Hatti-Kaul, R., Ulrika-Törnvall, U., Gustafsson, L. and Börjesson, P. (2007). Industrial biotechnology for the production of bio-based chemicals – a cradle-to-grave perspective. Trends Biotechnol. 25,119–124.

[18] Helgadóttir, A. and Kristjánsdóttir, T. (1991). Simple approach to the analysis of G×E interactions in a multilocational spaced plant trial with timothy. Euphytica 54,65–73.

[19] Kadkol, G.P., Halloran, G.M. and MacMillan, R.H. (1989). Shatter resistance in crop plants. Crit. Rev. Plant Sci. 8,169-188.

[20] Lin, C.S. and Binns, M.R., 1988. A superiority measure of cultivar performance for cultivar×location data. Can. J. Plant Sci. 68,193–198.

[21] Li, C.C. (1956). The concept of path coefficient and its impact on population genetics. Biometrics 12,190–210.

[22] Li, R., Yu, K., Hatanaka, T.and Hildebrand, D.F. (2010). Vernonia DGATs increase accumulation of epoxy fatty acids in oil. Plant Biotechnol. J. 8,184–195.

[23] Kang, M.S., Miller, J.D. and Tai, P.Y.P. (1983). Genetic and phenotypic path analyses and heritability in sugarcane. Crop Sci. 23,643–647.

[24] Mebrahtu, T., Gebremariam, T., Kidane, A. and Araia, W. (2009). Performance of *Vernonia galamensis* as a potential and viable industrial oil plant in Eritrea: yield and oil content. Afric. J. Biotechnol. 8,635-640.

[25] Mohamed, A., Mebrahtu, T. and Andebrhan, T. (1999). Variability in oil and vernolic acid contents in the new *Vernonia galamensis* collection from East Africa. *In*: Janick, J. (ed). Proc. Perspectives on New Crops and New Uses. ASHS. Press, Alexandria, VA. pp. 272–274.

[26] Morgan, C.L., Bruce, D.M., Child, R., Ladbrooke, Z.L. and Arthur, A.E. (1998). Genetic variation for pod shatter resistance among lines of oilseed rape developed from synthetic *B. napus*. Field Crops Res. 58,153-165.

[27] Morgan, C.L., Ladbrooke, Z.L., Bruce, D.M., Child, R. and Arthur, A.E. (2000). Breeding oilseed rape for pod shattering resistance. J. Agric. Sci. 135,347-359.

[28] Perdue, R.E. (1988). Systematic botany in the development of *Vernonia galamensis* as a new industrial oilseed crop for the semi-arid tropics. Symb. Bot. Ups. 24,125–135.

[29] Perdue, R.E., Carlson, K.D. and Gilbert, M.G. (1986). *Vernonia galamensis* potential new crop source of epoxy acid. Econ. Bot. 40, 54–68.

[30] Paux, E., Faure, S., Choulet, F., Roger, D., Gauthier, V., Martinant, J.P., Sourdille, P., Balfourier, F., Le Paslier, M.C., Chauveau, A., Cakir, M., Gandon, B. and Feuillet, C. (2010). Insertion site-based polymorphism markers open new perspectives for genome saturation and marker-assisted selection in wheat. Plant Biotechnol. J. 8,196–210.

[31] Ramalema, S.P., Shimelis, H, Ncube. I, Kunert, K.K. and Mashela, P.W. (2010). Genetic analysis among selected vernonia lines through seed oil content, fatty acids and RAPD DNA markers. Afric. J. Biotechnol. 8,117-122.

[32] Raman, H., Stodart, B., Ryan, P., Delhaize, E., Emberi, L., Raman, R., Coombes, N. and Milgate, A. (2010). Genome wide association analyses of common wheat (*Triticum aestivum* L) germplasm identifies multiple loci for aluminium resistance. Genome 53,957-966.

[33] Ray, D.T. (1994). Development of new oilseed crops in the USA. *In*: Hennink, S., van Soest, L.J.M., Pithan, K. and Hof, L. (eds.) Proc. Alternative Oilseed and Fibre Crops for Cool and Wet Regions of Europe, COST 814 Workshop. p. 26–33.

[34] Wageningen, Netherlands. 7–8 Apr. 1994. European Cooperation in the Field of Scientifi c and Technical Research.

[35] SAS Institute. (2004). Base SAS 9.1.3: Procedures guide. SAS Inst., Cary, NC.

[36] Sharma, H.C., Sharma, K.K., Seetharama, N. and Ortiz, R. (2001). Henetic transformation of crop plants: risks and opportunities for the rural poor. Curr. Sci. 80,1495-1508.

[37] Shimelis, H. and Hugo, A. (2011). Determination of selection criteria for seed yield and seed oil content in Vernonia (*Vernonia galamensis* variety *ethiopica*). Ind. Crops Prod. 33,436–439.

[38] Shimelis, H.A., Labuschagne, M.T. and Hugo, A. (2006). Variation of oil content and fatty acid compositions among selected lines of *Vernonia* (*Vernonia galamensis* variety *ethiopica*. S. Afric. J. Plant Soil 23,62–63.

[39] Shimelis, H., Mashela, P. and Hugo, A. (2008). Performance of *Vernonia* as an alternative industrial oil crop in Limpopo Province of South Africa. Crop Sci. 48,236-242.

[40] Slade, A.J., Fuerstenberg, S.I., Loeffler, D., Steine, M.N. and Facciotti, D. (2005). A reverse genetic, nontransgenic approach to wheat crop improvement by TILLING. Nature Biotechnol. 23,75-81.

[41] Tanksley, S.D., Young. N.D., Paterson, A.H. and Bonierbale, M.W. (1989). RFLP mapping in plant breeding: new tools for an old science. Biotechnol. 7,257-264.

[42] Thomas, R. (2003). Crop production in the Limpop Province. In: Nesamvuni, A.E., Oni, S.A, Odhiambo, J.J.O. and Nthakheni, N.D. (eds.) Agriculture as a cornerstone of the economy of the Limpopo Province: A study commissioned by the Economic Cluster of the Limpopo Provincial Government under the Leadership of the Department of Agriculture, Limpopo, Polokwane.

[43] Thompson, A.E., Dierig, D.A., Johnson, E.R., Dahlquist, G.H. and Kleiman, R. (1994a.) Germplasm development of *Vernonia galamensis* as a new industrial oilseed crop. Ind. Crops Prod. 3,185–200.

[44] Thompson, A.E., Dierig, D.A. and Kleiman, R. (1994b). Characterization of *Vernonia galamensis* germplasm for seed oil content, fatty acid composition, seed weight, and chromosome number. Ind. Crops Prod. 2,299–305.

[45] Wittkop, B., Snowdon, R.J. and Friedt, W. (2009). Status and perspectives of breeding for enhanced yield and quality of oilseed crops for Europe. Euphytica 170, 131-140.

[46] Wright, S., (1934). The method of path coefficients. Ann. Math. Stat. 5,161–215.

[47] Zhao, J., Paulo, M-J., Jamar, D., Lou, P., van Eeuwijk, F., Bonnema, G., Vreugdenhil, D. and Koornneef, M. (2007). Association mapping of leaf traits, flowering time, and phytate constent in *Brassica rapa*. Genome 50,963-973.

Chapter 6

Nutritional Management of Cereals Cropped Under Irrigation Conditions

Juan Hirzel and Pablo Undurraga

Additional information is available at the end of the chapter

1. Introduction

Crop nutritional management must be oriented so as to achieve economically convenient yields for the producer along with an efficient use of resources and concern for the environment. The use of information about soil chemical properties and experience of the behavior of each species under fertilization conditions with variations in soil chemical properties allow adjusting nutrient rates for different production situations. This chapter provides basic information about the nutritional requirements of the main cereals cultivated in the world, nutritional management strategies, and the nutritional value of using residues. This information is a guide for the producer to determine a nutritional management strategy using information provided by analyzing soil chemical properties for different productivity scenarios.

1.1. Corn

Corn (*Zea mays* L.) is a crop that can develop in a range of soil and climatic conditions [1]. It exhibits high nutrient extraction [2] and notably surpasses other crops such as small grain cereals and grain legumes. This cereal is grown for different purposes, but mainly for animal feed (silage and grain), poultry (grains), and pigs (grains), as well as for human consumption as grain, sweet corn, or corn.

Silage corn is an important food supplement in pastoral systems, particularly in the dairy industry where the main characteristics are high dry matter (DM) yield and metabolizable energy content [3]. Dairy cow producers in the central south and south in Chile use silage corn as feed between summer and winter, which allows extending lactation and increase production. This is done year-round in the central zone.

In the corn plant, the ratio DM grain:leaves + stem is considered as an important index of the nutritional value of the forage because of the high digestibility of the grain [3] as well as its starch content.

Nutrient concentration in the corn crop is highly variable just like in other crops and is associated with the genetic material used, the environment, and the agronomic management employed. [4] pointed out that the critical nitrogen (N) concentration in corn plants (aerial part) varied between 0.91 and 1.2% for corn with a development cycle that fluctuated between 141 and 154 d. [5] points out N concentrations between 0.72 and 0.80% for a semi-late corn hybrid in central Chile, which was fertilized with N rates similar to its extraction. For a similar experiment, this author indicates phosphorus (P) and potassium (K) concentrations that fluctuated from 0.10 to 0.11% and 0.60 to 0.64%, respectively, in fertilization conditions with increasing rates of these nutrients [6]. Meanwhile, [2] showed nutritional fluctuations between 0.57 and 1.15% for N; 0.13 and 0.23% for P; 0.74 and 1.2% for K; 0.18 and 0.40% for calcium (Ca); and 0.10 and 0.17% for magnesium (Mg) in two mid-season silage hybrids cultivated during two consecutive seasons with different fertilization treatments in central south Chile. The same study includes the high variability found in the N, P, Ca, and Mg concentrations between both genetic materials used as has also been corroborated in other fertilization experiments conducted in the same crop.

For sweet corn, macronutrient concentrations in ears of eight hybrids in the harvest stage evaluated in the United States corresponded to 1.35-2.10% for N; 0.24-0.32% for P; 0.98-1.24% for K; 0.037-0.083% for Ca; 0.10-0.15 for Mg; and 0.09-0.15 for sulfur (S), whereas concentrations of these nutrients in the aerial residue corresponded to 1.96-2.49 for N; 0.19-0.27% for P; 2.35-3.26% for K; 0.33-0.41% for Ca; 0.21-0.28% for Mg; and 0.18-0.22% for S [7].

For grain corn, fertilization experiments carried out by Instituto de Investigaciones Agropecuarias (INIA) in central Chile indicate N concentrations from 0.77 to 1.18%; P concentrations from 0.13 to 0.17%; and K concentrations from 0.29 to 0.30% [5,6]. Meanwhile, experiments conducted on different commercial hybrids in central south Chile (2003 to 2010) indicate that macronutrient concentration in the grains at the harvest stage fluctuated between 1.13 and 1.77% for N; 0.31 and 0.52 for P; 0.37 and 0.58 for K; 0.010 and 0.037 for Ca; 0.14 and 0.19 for Mg; and 0.11 and 0.15 for S (n = 160 samples).

While there are many commercial hybrids with differences in earliness in their crop cycle, in general all of them exhibit a marked extraction of nutrients that is accentuated at the sixth leaf as shown in Figure 1.

It should also be noted that the high K extraction in this crop, which is expressed as K_2O, is far superior to the N requirements, which also occurs in many plant species.

As for nutrient extraction carried out only for corn grain, [1] point out N:P:K extractions of approximately 16:3:3.3 for each ton of grain, respectively. [5,6] indicates N:P:K extractions with means of 9:1.4:2.5 for each ton of grain, respectively. The differences identified by these authors are due to the abovementioned variations in the genetic materials used.

Experimental records generated by fertilization studies of the corn crop for grain and silage allow estimating the nutritional requirements shown in Table 1. Requirements in this table are

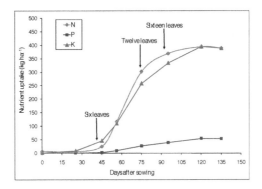

Figure 1. Nutrient extraction in corn crop [8].

shown as a function of the yield unit. Nutritional requirements in Table 1 are highly variable and this variability depends on many factors among which are genetic material, soil physico-chemical properties (nutrient availability in the environment), and also climatic conditions as pointed out by [4-6,9].

Nutrient	Type of corn and nutritional requirement	
	Grain (kg Mg⁻¹)	Silage (kg Mg⁻¹ MS)
N	14 – 26	6.9 – 14.5
P_2O_5	6 – 13	3.1 – 6.9
K_2O	16 – 38	8.2 – 23.2
CaO	3.0 – 7.5	1.5 – 4.5
MgO	3.2 – 7.4	1.7 – 4.0
S	1.4 – 2.6	0.8 – 1.3
Fe	0.24 – 0.41	0.109 – 0.185
Mn	0.04 – 0.06	0.019 – 0.029
Zn	0.03 – 0.05	0.013 – 0.021
Cu	0.002 – 0.009	0.001 – 0.004
B	0.016 – 0.018	0.007 – 0.008

Source: Adapted from [10].

Table 1. Nutritional requirements of rice crop for grain or silage expressed as a function of the yield unit (n = 240 samples of six commercial hybrids).

Nutrient distribution in the corn plant is shown in Figure 2. This figure shows that a large part of N and P extracted by the corn plant is concentrated in the grain, whereas the aerial residue concentrates an important part of K and Mg absorbed by the plant, this was also pointed out by [5,6]. In turn, Ca is mainly concentrated in the aerial residue. Therefore, when incorporating corn residues, a large part of K, Ca, and Mg are recycled, which contributes to reducing the requirements of these nutrients in the next crop to the extent that the incorporated residue achieves its biological decomposition in the soil.

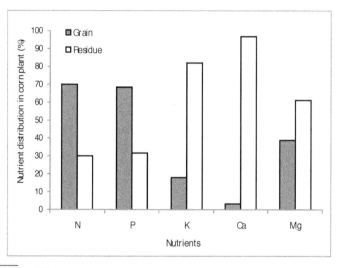

Source: Adapted from [10].

Figure 2. Nutrient distribution in corn plant.

Regarding the effect of applying different nutrients to the crop, various authors have demonstrated positive effects of applying N, P, K, Ca, and Zn [4,5,10-16]. The favorable effect of organic amendments as a fertilizer source has been demonstrated on grain and dry matter yield for silage using fresh amendments (for example, poultry and swine litter) and composted manures (pig compost, dairy compost, and others), leaving a positive residual effect on nutrient availability for the next crop [2,10,17-24].

Another management factor to consider in corn crop fertilization is residue management. Some theoretical models indicate that the N contained in the residues is gradually mineralized by the soil microbial biomass once it has been incorporated into the soil and is available for the next crop. However, this availability has not been evidenced under experimental field conditions since residue decomposition is also due to initial fractionation thereof (trituration) and soil temperature. Furthermore, soil biomass tends to generate humic compounds [5], which exhibit a C/N ratio that fluctuated between 18 and 22 [25]; therefore, an important fraction of N contained in the residue will be part of the humic compounds that will be synthesized by the soil's biological activity, situation that allows increasing soil organic matter content and improve its physical, chemical, and biological properties.

[26] indicate that superficially applying three residue levels (including a control without residues) in a corn grain monocrop fertilized with increasing N rates (from 0 to 225 kg ha^{-1}) through various fertilization sources, reduced soil inorganic N availability with a slightly negative effect on yield as compared with the treatment with no residues. In addition, soil temperatures were negatively affected by the increasing residue rate. For evaluated N sources, the best effect was achieved by ammonium nitrate over the use of urea.

[27] indicate that using three residue levels (including a control without residues) in a corn grain crop managed with two tillage systems (chisel and no-till) and two increasing N rates (between 67 and 268 kg ha^{-1}) exhibited a variable effect on yield according to the rainfall level and tillage practice used. Thus, for low rainfall and tillage conditions, total residue removal (control) produced a higher grain yield than using residues at the highest rate. Residues applied at high rates with no-till for this low rainfall level had a positive effect on yield. In summary, for normal rainfall conditions, the level of residues used in tillage conditions had no effect on yield, while for no-till management, the total (control) or partial removal of residues produced a higher grain yield than using residues at the highest level.

1.2. Wheat

This crop is one of the three most important cereals for human consumption along with corn and rice. History indicates that wheat (*Triticum aestivum* L.) was first cultivated by ancient hunter-gatherers in Southwest Asia, and archeological remains have been found of bread wheat from Turkestan in 6000 B.C. It has been established that the first domesticated wheat in the world goes back to 7500 to 6500 B.C. The wheat crop, along with other cereals, was first introduced to Chile by Pedro de Valdivia in 1540 and since then it has become the most widespread crop in Chile [28]. In nutritional terms, the wheat crop is characterized by its high N and K requirement as well as other essential nutrients such as P, S, and Ca [29,30]. In general, N and K represent about 80% of total nutrients in wheat plants; together P, S, Ca, and Mg make up 19%, while total micronutrients are less than 1% [31].

Wheat plants absorb the N nutrient as nitrate ion or ammonium ion. The way in which N translocation occurs depends on the absorbed N source and root metabolism [9]. Absorbed N is transported by the xylem to the leaves as nitrate ion or could be reduced in the roots and transported in inorganic form as amino acids or amides. A large part of the absorbed ammonium has to be incorporated in organic compounds in the roots [32]. In terms of phloem, N is a mobile nutrient, thus under deficiency conditions this element can be retranslocated from the older to the younger leaves and then translocated from there to the developing grains. The principal forms of organic N in the phloem sap are amides, amino acids, and ureides [33]. The nitrate and ammonium ions are not present in this sap, but mainly in the xylem. Nitrogen in the wheat crop is the major component of proteins, amino acids, enzymes, and nucleic acid [9]. It is also part of the mature grain, mainly concentrated as proteins in the endosperm which is the part that makes up the flour [28].

A deficiency of N in plants greatly reduces growth rate. In the case of cereals [9], tillering is poor and the leaf area is small; both the number of spikes per unit of area and the number of grains per spike are reduced. Since this nutrient is a component of chlorophyll, its deficiency is seen as a generalized yellowing or chlorosis of the leaves, appearing first on the lower leaves while the higher leaves remain green. In cases of severe deficiency, a generalized chlorosis is noticeable in the whole plant. Finally, it decreases crop yield and the grain protein content [33]. Excesses of N are less evident than its deficiency. They include prolonged plant growth, dark green coloration of foliage, increased plant susceptibility to the attack of phytopathogens, and a delay in crop maturity.

The concentration of N in the wheat plant decreases over the phenological periods reaching values of 3.5-4.2% at the full tillering stage to 0.9-1.2% at the harvest maturity stage [34]. The adequate concentration in the higher leaves before the spike formation stage fluctuates between 1.75 and 3.0% [29]. In the case of wheat grain, N concentration fluctuates between 1.6 and 2.4% at harvest and surpasses other nutrients. Phosphorus deficiency restricts plant development, delays growth, tillering, root development, and maturity. Deficiency symptoms normally start in the oldest leaves and are characterized by a blue-greenish to reddish color, which can lead to a reddish color and bronze tints that normally start from the edges. Leaves often have a darker green color than normal plants. This is because the expansion of cells and leaves is delayed more than chlorophyll formation, so that the chlorophyll content per leaf area unit is higher [32]. A symptom of P deficiency is the decrease of the stem/root ratio and less growth of all the growth points. Extremely high P levels can result in toxicity symptoms, which generally occur as aqueous points in the leaf tissue, eventually becoming necrotic [33]. In very severe cases, P toxicity can provoke plant death.

Wheat plant P concentration decreases with the maturity process and can vary between 0.23 and 0.30% at the full tillering stage and decrease to values of 0.12-0.18% at harvest maturity stage [34].

Potassium deficiency is produced as chlorosis along the edge of the leaf followed by burning and bronzing of the old leaf tips. The affected area is curled when the deficiency of this element increases. The symptoms of K deficiency appear in the old leaves due to nutrient mobility. The affected plants are generally stunted and have shortened internodes. These plants exhibit slow and stunted growth, weak culms susceptible to lodging, higher incidence of pests and diseases, lower yields, curled grains, and low grain quality [33]. Plants with K deficiency can lose the control of the respiration rate and exhibit internal water deficit. High K concentrations contribute to increasing plant tolerance and resistance to diseases and pests.

Wheat plant K concentration decreases with crop maturity, fluctuates from 3.8-4.5% at the full tillering stage, and decreases to 0.9-1.2% at the harvest maturity stage [34].

Calcium deficiency produces small, twisted, dark green leaves [33]. Although all the growth points are sensitive to Ca deficiency, the root meristems are the most affected. Calcium is a non-toxic mineral nutrient even in high concentrations and is very effective to detoxify high concentrations of other mineral elements in plants [32]. Moreover, high Ca contents within the plant raise tolerance and resistance against diseases and pests.

Wheat plant Ca concentration decreases with maturity and reaches values of 0.28-0.30% at the full tillering stage to levels of 0.08-0.10% during harvest maturity.

Magnesium is an element that easily translocates from the older parts to the younger parts; therefore, its deficiency symptoms first appear in the oldest parts of the plant. A typical symptom of Mg deficiency is interveinal chlorosis of the old leaves in which the veins remain green but the area between them turns yellow. When the deficiency becomes more severe, leaf tissue uniformly turns chlorotic, then brown and necrotic. Leaves are small and break easily [33].

Wheat plant Mg concentration tends to decrease with maturity [34] and can fluctuate from 0.14-0.16% at the full tillering stage to values of 0.05-0.07% at harvest maturity.

Common Zn deficiency symptoms in wheat are arrested plant growth, poor tillering, light green coloring, yellowing, whitened spots, chlorotic stripes on both sides of the central vein, and small leaves. The internodes are short and the flowering, fructification, and maturity processes can be delayed. A high soil P concentration can induce Zn deficiency [9]. The toxicity of Zn can be translated in a reduction of root growth and leaf expansion followed by chlorosis. This is associated with concentrations higher than 200 mg kg^{-1} Zn in the tissue. The excess of Zn can induce Fe deficiency, which is recognized by interveinal chlorosis in the new leaves of the plant.

Wheat plant Zn concentration decreases with maturity and reaches levels of 12 to 20 mg kg^{-1} at the full tillering stage and 10 to 12 mg kg^{-1} at cereal harvest maturity [34].

Regarding the nutritional management of the wheat crop, it must be considered that the higher the yield level, greater the requirement of nutrients for the crop will be. However, in the case of the wheat crop, it is fundamental to also consider the cultivar (commercial variety) and the growth habit. For the cultivar, differences are observed in the accumulation of proteins in the grain, thus indicating variations in plant N requirement. Each cultivar exhibits a unique genetic base that gives it unique attributes for potential yield and grain quality [35]; [36]. For growth habit, winter habit cultivars generally produce higher total nutrient extractions than those of spring habit, mainly because it is more permanent in the soil. Other factors affecting variability in the nutritional requirements of the wheat crop are soil physicochemical properties, climatic conditions, and agronomic management. In this way, the same wheat cultivar sown at different sites will also exhibit different nutritional requirements for the same yield levels. Therefore, to define nutrient quantities to apply to this crop and in other crops, factors that influence these nutritional requirements must be considered.

Wheat crop nutritional requirements can be observed in Figure 3 with nutrient extraction in wheat cv. Dollinco-INIA (cultivar with alternative habit) showing gradual extraction that reaches its maximum near the end of the cultivation period, and where N is the most extracted nutrient. Nutrient extraction of this cultivar in its maximum accumulation period is 30.6; 5.7; 24.7; 7.9; and 4.1 kg Mg^{-1} N, P$_2$O$_5$, K$_2$O, CaO, and MgO, respectively, for a grain yield of 8 Mg ha^{-1}. For the highest extracted nutrient, other cultivars exhibit a higher K extraction, for example, the cvs. Tukán-INIA and Domo-INIA with nutrient extraction of their maximum accumulation period is 26.4-31.1; 6.0-7.4; 30.7-32.5; 4.0-8.6; and 2.1-4.2 kg Mg^{-1} N, P$_2$O$_5$, K$_2$O, CaO, and MgO, respectively, for a grain yield of 7 and 9 Mg ha^{-1} for each cultivar, respectively [31].

The nutrient extraction curves allow adjusting and harmonizing better the fertilization criteria with the growth rate of the wheat crop. For example, in the initial phenological stages of cv.

Kumpa-INIA (Table 2) a large part of the extraction of evaluated nutrients is produced. Thus, nearing the end of the vegetative stage (flag leaf visible) there has been an extraction of 80.3% N; 76.6% P; 98.4% K; 62.6% Ca; and 67.1% Mg, while for cv. Dollinco-INIA (Table 3) nearing the end of the vegetative stage (flag leaf visible) there has been an extraction of 56.9% N; 33.9% P; 62.2% K; 53.8% Ca; and 38.4% Mg. However, the accumulated absorption values are less than those of cv. Kumpa-INIA [31].

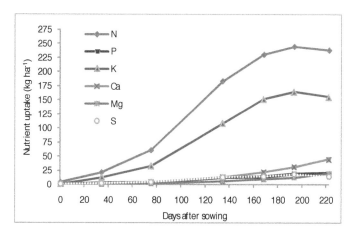

Figure 3. Seasonal nutrient uptake of winter wheat crop cv Dollinco-INIA.

Phenological stage	Days after sowing	Accumulated percentages[1]				
		N	P	K	Ca	Mg
End of tillering	111	27.3	16.1	36.2	18.3	15.4
Two nodes	133	46.3	38.6	51.0	33.5	30.1
Flag leaf visible	153	80.3	76.6	88.4	62.6	67.1
Emergence	174	98.5	95.9	97.2	93.1	95.1
Anthesis	182	100.0	100.0	100.0	100.0	100.0

[1]Maximum extraction obtained during the evaluated crop development cycle corresponded to 269 kg N, 26 kg P, 249 kg K, 36 kg Ca, and 14 kg Mg.

Table 2. Percentage extraction of N, P, K, Ca, and Mg in winter wheat cv. Kumpa-INIA.

As for the effect of applying different nutrients to the crop, various authors have demonstrated positive effects of applying N, P, Zn, and liming [28,31,37-41]. For wheat crop nutrient partialization many experimental studies have demonstrated that there is a response to

Phenological stage	Days after sowing	Accumulated percentages[1]				
		N	P	K	Ca	Mg
End of tillering	118	27.1	11.4	29.6	13.9	10.6
Two nodes	140	43.6	30.5	53.1	35.7	28.8
Flag leaf visible	160	56.9	33.9	62.2	53.8	38.4
Emergence	167	98.9	89.5	100.0	100.0	100.0
Anthesis	181	100.0	100.0	100.0	100.0	100.0

[1]Maximum extraction obtained during the evaluated crop development cycle corresponded to 286 kg N, 33 kg P, 244 kg K, 36 kg Ca, and 16 kg Mg.

Table 3. Percentage extraction of N, P, K, Ca, and Mg in winter wheat cv. Dollinco-INIA.

partially applying N and that yield is generally maximized with three applications of this nutrient, which correspond to sowing (15-20% total N), start of tillering (40-50% total N), and start of nodes (30-40% total N) [28,31,41]. Experimental studies indicate that there are differences among cultivars and soil and climatic conditions for the rate of this nutrient, but that yield is generally maximized when the N rate is determined based on the absorption requirement of the crop (replacement rate) with a variation that can fluctuate between 90 and 120% of crop consumption [31,34]. The rates of the other nutrients must be adjusted to the nutritional requirements and soil chemical properties.

The incorporation of residues of the wheat crop is a practice that benefits physical, chemical, and biological soil properties. Figure 4 shows that an important fraction of N, P, and Mg along with the greater part of K and Ca extracted with this crop will be returned to the soil along with the incorporation of residues, which will have a significant effect on the reduction rate of some of these nutrients in the next crop. However, the required application of N should be considered to achieve an adequate decomposition of the incorporated residue so as not to affect N availability for the next crop.

1.3. Rice

Rice (*Oryza sativa* L.) is one of the most important cereals for development in the world and a basic food for at least half of its population. Generally, an annual semi-aquatic crop is considered and more than 20 species of the *Oryza* genus are recognized of which only two are cultivated [42,43], which can be aquatic, semi-aquatic, and dry land.

Among the cereals produced on a world level, rice occupies the greatest proportion of soils. Of the 147.5 million ha of soil dedicated to rice production in the world in 1989, developing countries contributed with 141.4 million ha, that is, 96% [44]. In the 1976-1980 period, a mean of 37.842 ha were sown [45], decreasing to 22.733 ha in the 2006-2009 period [46], which indicates a decrease of approximately 40% of the area in 25 years.

As for the nutritional functions and effects of the nutrients in the plant, N must be constantly supplied to the crop to achieve an adequate harvest, especially during panicle formation and

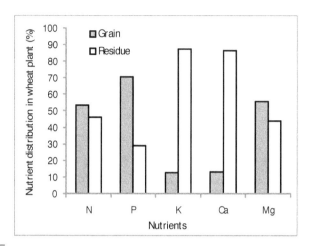

Source: [34].

Figure 4. Nutrient distribution in bread wheat plant (mean of cvs. Domo-INIA and Quelén-INIA).

development. The final yield of the crop is defined by the number of panicles per m^2 and number of tillers, which is defined within the first 10 d after maximum tillering. These components are influenced by N availability during these development stages [47,48]. Meanwhile, the number of spikelets per panicle is closely related to the N content of the leaf sheath during the weeks prior to flowering [47,49]. Nitrogen will later influence an efficient assimilation of carbohydrates on the part of the grain and in a correct filling thereof. The form of absorbed N during the first growth stages the rice plant prefers the ammonium forms (NH_4^+) [50,53], while in the stages near maturity it prefers N in nitrate form (NO_3^-). Nitrogen as ammonium is favorable up to the panicle initiation stage [51], and later nitrate absorption is promoted, especially during the panicle embryonic formation stage, stimulating the increase in the number of panicle flowers and grain weight [52,53]. Ammonium N increases the number of tillers and the number of panicles per plant [52]. Although in flooded soil conditions it is preferred to apply ammonium forms of N or those derived from ammonia (urea), depending on tilling depth, in the first 5-10 cm of soil an aerobic condition appears which would allow transforming ammonium to nitrate, a process that is accelerated by temperature accumulation. Notwithstanding the above, there is no scientific evidence in Chile about the modern varieties that indicates the advantages of applying N in the nitrate form after sowing. Nitrogen excesses, especially associated to abnormal climatic phenomena during the reproductive stage, particularly low temperatures [53], produce growth and internal distribution disorders of this nutrient that promotes vegetative growth and can delay harvest and reduce yield.

Internal distribution of N in the rice plant is established in studies conducted with N^{15} and which indicate that between 3 and 5% is found in the roots, between 30 and 44% is found in the aerial vegetative structures, between 37 and 40% is found in the panicle, and between 7 and 13% is found in the senescent structures [54].

Phosphorus is an element that promotes root growth; therefore, it improves the absorption of water and other nutrients. It also increases the resistance to periods of reduced water availability. An adequate supply of P improves grain flowering and fertilization, increases crop precocity, and increases 1000 grain weight [52]. The highest P absorption occurs during the vegetative development period and decreases after the panicle embryonic formation [53]. In turn, this absorption is promoted by temperature increases with an optimum temperature of 30 °C. Availability of P in flooded soils is promoted by the redox processes and liberating part of the P retained as iron phosphate and aluminum [55].

Potassium plays an important role at the beginning of vegetative growth with an emphasis during tiller formation and has a direct influence on the determination of the final number of panicles. During the ripening stage it promotes synthesis and translocation of low molecular weight carbohydrates, is involved in activating phosphorylation processes (energy transport) to activate the transport of soluble N compounds to grains in formation, avoids their accumulation in other tissues, and also promotes 1000 grain weight [53,56]. Furthermore, K increases plant resistance to various diseases, such as stem rot detected in Chile [57], and adverse climatic conditions (high levels of solar radiation and temperature, or low temperature during tillering and flowering), playing an important role in the economy of use and loss of water by transpiration from the plant [53].

Moreover, K stimulates cell division, is involved in photon transport in photosynthesis, directs the synthesis of starch, inulin, amino acids, and proteins, modifies cell permeability, interferes with plasmolysis and turgidity mechanisms, and together with P and Mg is actively involved in carbohydrate metabolism [53].

Sulfur is an important element during the whole growth cycle of the rice plant; it strongly affects grain quality since it is part of some essential amino acids and part of N metabolism in synthesizing proteins and carbohydrates. It reduces nitrate, catalyzed chlorophyll formation along with Cu and Fe, acts as a hydrogen carrier, helps to regulate the tricarboxylic acid cycle, and is part of the sulfur radicals (SH) [53]. The quantities of S absorbed by the crop are relatively low as compared to other nutrients (N and K) since they are covered by the soil's natural contribution (organic matter mineralization), therefore, it is uncommon to apply S fertilizers in the rice crop. The availability of S is reduced in flooded systems as a result of redox reactions [55].

Calcium contributes to plant rigidity and resistance to lodging. Magnesium is located in the pyrrole rings making up chlorophyll and is a catalyst in the enzyme activity of nitrate reductases or self-induced enzymes that require molybdenum (Mo) [53]. The availability of Ca and Mg in flooded systems is improved as a result of redox processes [55].

Another important element for the rice crop is silicon (Si). Silicon extraction by the rice plant is higher than for any other mineral element with a concentration that can fluctuate between 2 and 9% of the plant dry matter [53]. This element is mainly deposited in the leaf epidermal cells and forms a siliceous double layer which is responsible for disease resistance. Studies conducted in Japan point out an extraction mean of 433 kg Si ha^{-1}. In quantitative terms, for each ton of paddy grain yield, the rice plant extracts 100 kg Si, of which a large part is concentrated in the rice husk [52].

Silicon is involved in the whole growth cycle of the rice plant, mainly affecting the stage between panicle formation and grain ripening. This essential element for rice cultivation promotes length development and oxidative activity of the root system in addition to protecting plants from Fe and Mn toxicity produced in anaerobic conditions of flooded soil. Finally, a good soil Si level improves P availability for the rice crop. With regard to the micronutrients, Fe participates in chlorophyll formation, prevents chlorosis, and is part of enzyme activity. The excess of Fe can inhibit K absorption [53]. Boron contributes to N uptake, participates in Ca metabolism, and stimulates meristematic activity and pollen formation [53]. Zinc stimulates initial plant development and its deficiency can affect potential crop yield. In flooded systems and as a result of redox processes, availability of Fe, Mn, and Mo increases and availability of Zn and Cu decreases [55].

Figure 5 shows the nutritional requirements of the rice crop. It can be observed that there is a gradual nutrient extraction in cv. Diamante-INIA (main cultivar used in Chile) between the initial and maximum tillering stages followed by slight increases in N and K accumulation unlike other nutrients that have an increased accumulation until later crop stages. In turn, the nutrient with the highest extraction is K. In productive terms, the nutritional requirements are 12 kg N; 8.4 kg P_2O_5; 22 kg K_2O; 6.2 kg CaO; and 5.6 kg MgO for each ton of yield. As additional information and as a guide, micronutrient requirements for this cultivar are 6.4 kg Fe; 1.88 kg Mn; 56 g Zn; 12 B; and 15 g Cu for each ton of yield.

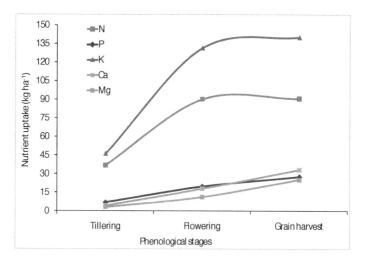

Figure 5. Seasonal nutrient uptake of rice crop cv Diamante-INIA for a 7.5 Mg ha⁻¹ yield.

Although the nutritional requirements of the rice crop are high for some nutrients such as K, the harvested grain from the field is usually extracted and the residue left in place to be incorporated or burned in the following soil preparation. Incorporating residues allows replacing much of the K, Ca, S, and microelements: however, when residues are burned prior to the following crop, there is a loss of N and S content in the residues through the volatilization process to the atmosphere.

Figure 6 shows the nutrient distribution in the rice plant and considers the grain and the residue; it estimates the nutritional contributions produced by incorporating residues and establishes the importance of carrying out this task. This figure shows that incorporating rice residues to the soil returns a large part of K, Ca, and Mg extracted from the previous rice crop and greatly reduces the nutritional requirements of the following crop.

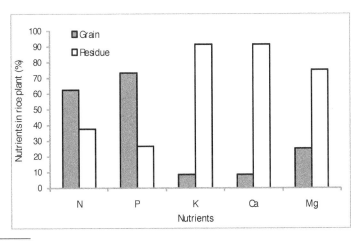

Adapted from [58].

Figure 6. Nutrient distribution in rice plant (grain and aerial residue).

Nutrient management of the rice crop should pay greater attention of the N rate and its form of partialization. The rate must be determined as a function of crop productivity that generates differences in N requirement and the ability of natural soil contribution (with differences between different soil taxonomic orders associated with the evolution of secondary materials and their relationship with active pools of organic matter) since N absorption is mainly from reserves (organic matter mineralization, microbial biomass dynamic cycles, and $N-NH_4+$ fixed in clays) [59-62], N fertilization [63], as well as a small fraction derived from irrigation water and other environmental and biotic sources. The ability of is determined through mineralization in conditions that are similar to field conditions [64-68]. To determine this soil natural N supply ability, the samples must be incubated under optimal mineralization conditions; flooded soils with soil:water ratio of 1:2.5, a temperature from 20ºC to 40ºC for a period of 14 to 21 d without agitating the samples [69]. Using N rates that are higher than the crop requirements can generate phenological development disorders, increasing the vegetative period and producing the start of the reproductive period at a later time, which can increase flower sterility due to low temperatures that can occur in this period since in the final stage of the crop environmental temperatures start decreasing, negatively affecting grain ripening and ultimately commercial yield. It is common to find situations in which production has not achieved adequate grain maturity associated with the incorrect N rate (rates higher than necessary for the particular edaphoclimatic situation). Furthermore, when crop development time is reduced, it also decreases yield potential and thus N requirements.

About N partialization, [70] indicated that the highest grain yield is produced when applying N in 2 or 3 phenological stages; (1) 50% at sowing and 50% at panicle initiation, or (2) 33% at sowing, 33% at tillering initiation, and 34% at panicle initiation. This can be observed in Figures 7 and 8. The partialization strategy that the producer must use will depend on the opportunity to carry out the tasks (sowing date and its relationship with the adequate date) and the area sown.

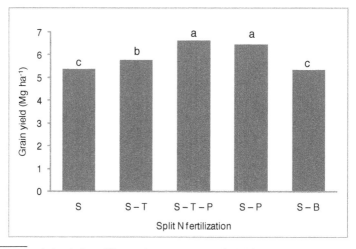

Different letters over the bars indicate differences between treatments (p <0.05)
S = 100% N at sowing; S – T = 50% of N at sowing and 50% at tillering; S – T – P = 33% N at sowing, 33% at tillering, and 34% at panicle initiation; S – P = 50% N at sowing and 50% at panicle initiation; S – B = 50% N at sowing and 50% at boot stage.

Figure 7. Grain yield of rice with different N application times on a vertisol soil in the Chilean central south region with pooled date for 2007-08 and 2008-09 seasons.

Experiments conducted in Chile with other nutrients have reported a response when applying K, Zn, B, and liming [71]. Rates of P, K, Mg, S, B, and Zn in the rice crop must be adjusted to nutrient requirement (as a function of yield) and the level of availability of each element in the soil. Meanwhile, Ca contributions will be carried out periodically to the extent that liming is performed.

Regarding alternative fertilizers, using organic amendments such as poultry litter has produced crop yield increases as compared to conventional fertilizers with equal N rates [22], which responds to the accumulative loss of soil organic matter with continuous crop cycles. Adding organic matter to the soil, including as crop residues, modifies the redox potential of the soil system and produces higher availability of some nutrients [72].

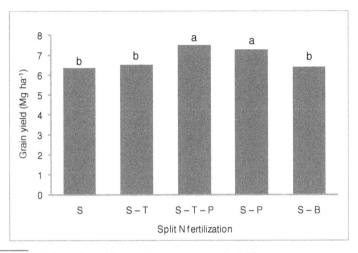

Different letters over the bars indicate differences between treatments (p <0.05)
S = 100% N at sowing; S – T = 50% N at sowing and 50% at tillering; S – T – P = 33% N at sowing, 33% at tillering, and 34% at panicle initiation; S – P = 50% N at sowing and 50% at panicle initiation; S – B = 50% N at sowing and 50% at boot stage.

Figure 8. Grain yield of rice with different N application times on an inceptisol soil in the Chilean central south region with pooled date for 2007-08 and 2008-09 seasons.

Finally, incorporating residues in the rice crop recycles a large part of the absorbed nutrients in the previous crop cycle [58] as can be observed in Figure 6 and reducing at the same time the rates of elements such as P, K, Ca, and Mg.

2. Nutritional management of main cereals cultivated as a function of yield and soil chemical properties

As abovementioned, the nutritional requirement of a crop is determined by its productivity, whereas the nutrient rate to be applied to the crop must be contrasted with the soil natural nutrient supply ability.

In a conceptual model for the same yield unit (Mg or other), the nutritional requirement can be expressed as a function of this yield unit and then be associated with the yield level as expressed in equation (1):

$$\text{Nutrient rate}\left(\text{kg ha}^{-1}\right) = \text{Nutritional requirement}\left(\text{kg Mg}^{-1}\right) * \text{Yield (Mg ha}^{-1}) \qquad (1)$$

The nutritional requirement per yield unit is inversely proportional to the natural nutrient supply ability determined by the soil as shown in the conceptual model in Figure 9. Thus, to the extent that the nutrient concentration at issue exhibits a higher level in the soil, the fertilization rate will be lower per yield unit to produce. In turn, there is a minimum and maximum rate for each nutrient, which means that when the nutrient concentration in the soil is less than the minimum value, the maximum rate is applied, whereas when the concentration of this nutrient in the soil is greater than the maximum value shown in the range (minimum rate per yield unit), the minimum rate will be maintained, which for some nutrients could be zero (not applied).

The nutrient rate in corn, wheat and rice crops is calculated in function of production level, expressed in Megagrams per hectare. In corn crop are utilized the figures 10, 11, 12, 13, 14, and 15 for N, P2O5, K2O, CaO, MgO, and S, respectively. For wheat are utilized the figures 16, 17, 18, 19, 20, and 21. And for rice crop are utilized figures 22, 23, 24, 25, 26, and 27, respectively.

Another variable to consider in the fertilization program is the correction soil acidity, for which liming is used by applying a rate determined by specific analyses in each soil and where it is determined if calcium carbonate ($CaCO_3$) is required to increase soil pH (acidity or alkalinity) in a given quantity. When liming is applied, the fixed Ca application is suspended by analyzing the soil. Likewise, if the amendment has to use Mg ($CaCO_3$*$MgCO_3$), the application of fixed Ca and Mg is suspended by analyzing the soil. In general, it is considered that when soil pH is greater than 6.0 liming is not necessary since there are no acidity problems (negative effect of aluminum (Al) on plant development under acid pH conditions, unless a nutritional imbalance in the soil that affects Ca and/or Mg needs to be corrected.

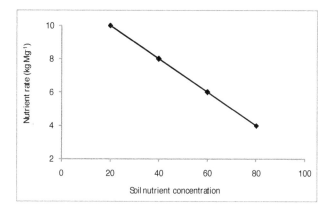

Figure 9. Nutrient rate ratio to apply to a crop (kg per yield unit) as a function of its concentration in the soil.

For the elements K, Ca, and Mg, which participate in cation exchange capacity of the soil (CEC) and given the phenomenon of competition or antagonism, the rate should be adjusted as a function of de percentage participation of each one of them in the CEC of the soil, considering that the element that must exhibit the highest participation in CEC is Ca, followed by Mg, and then K. Thus, the reference ranges of the saturation percentage of each one of these elements are shown in Table 4.

Element	Deficient range	Adequate range	High range
	— — — — — — — — — % over CEC — — — — — — — —		
K	< 2	2 – 3	"/>3
Ca	< 55	55 - 65	"/>65
Mg	< 10	10 – 15	"/>15

Table 4. Reference ranges for K, Ca, and Mg saturation levels in the cation exchange capacity of the soil (CEC)

2.1. Nutrient rates in corn crop

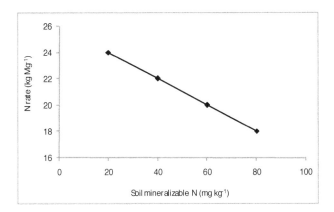

Figure 10. N rate to apply to corn crop for each yield unit as a function of soil mineralizable N concentration.

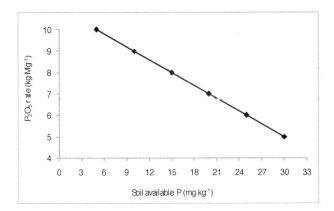

Figure 11. P$_2$O$_5$ rate to apply to corn crop for each yield unit as a function of soil available P concentration (Olsen method).

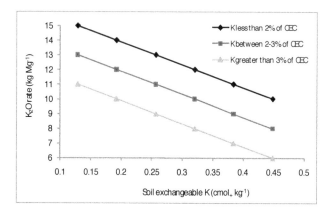

Figure 12. K₂O rate to apply to corn crop for each yield unit as a function of soil exchangeable K concentration.

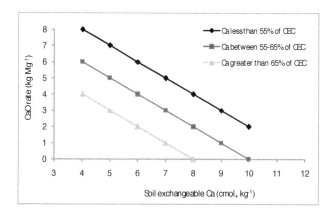

Figure 13. CaO rate to apply to corn crop for each yield unit as a function of soil exchangeable Ca concentration.

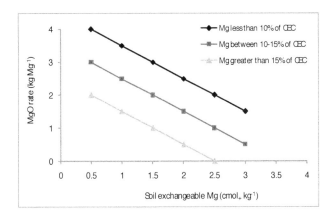

Figure 14. MgO rate to apply to corn crop for each yield unit as a function of soil exchangeable Mg concentration.

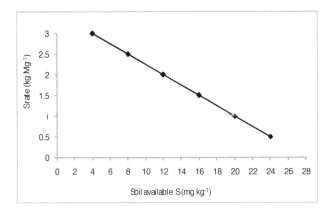

Figure 15. S rate to apply to corn crop for each yield unit as a function of soil available S concentration.

2.2. Nutrient rates in wheat crop

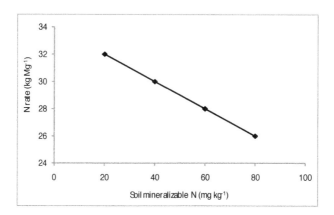

Figure 16. N rate to apply to wheat crop for each yield unit as a function of soil mineralizable N concentration.

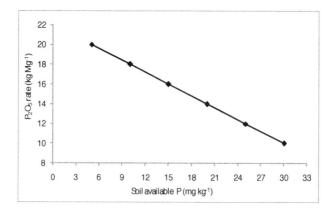

Figure 17. P_2O_5 rate to apply to wheat crop for each yield unit as a function of soil available P concentration (Olsen method).

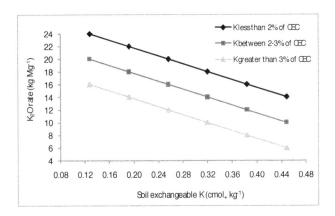

Figure 18. K$_2$O rate to apply to the wheat crop for each yield unit as a function of soil exchangeable K concentration.

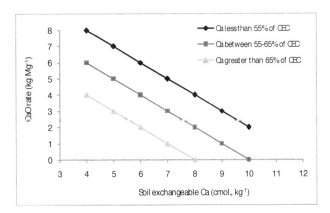

Figure 19. CaO rate to apply to wheat crop for each yield unit as a function of soil exchangeable Ca concentration.

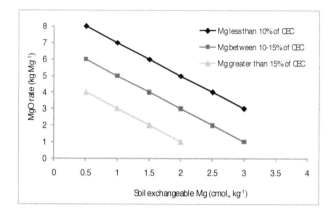

Figure 20. MgO rate to apply to wheat crop for each yield unit as a function of soil exchangeable Mg concentration.

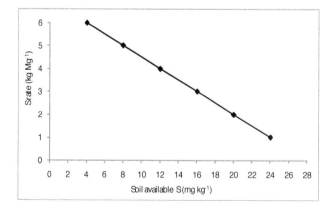

Figure 21. S rate to apply to wheat crop for each yield unit as a function of soil available S concentration.

2.3. Nutrient rates in rice crop

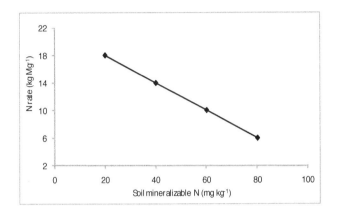

Figure 22. N rate to apply to rice crop for each yield unit as a function of soil mineralizable N concentration.

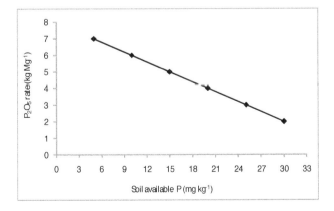

Figure 23. P_2O_5 rate to apply to rice crop for each yield unit as a function of soil available P concentration (Olsen method).

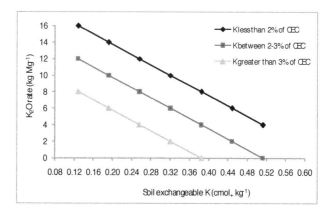

Figure 24. K₂O rate to apply to rice crop for each yield unit as a function of soil exchangeable K concentration.

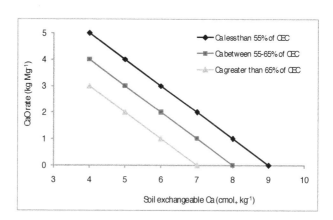

Figure 25. CaO rate to apply to rice crop for each yield unit as a function of soil exchangeable Ca concentration.

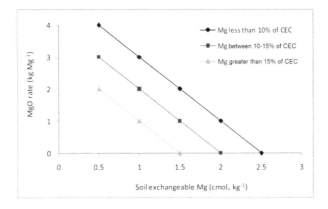

Figure 26. MgO rates to apply to rice crop for each yield unit as a function of soil exchangeable Mg concentration.

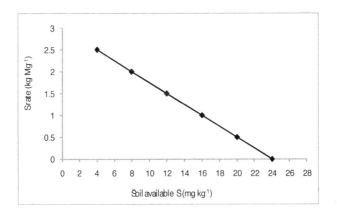

Figure 27. S rate to apply to rice crop for each yield unit as a function of soil available S concentration.

As an example of applying this methodology, sowing corn with an expected yield of 14 Mg ha^{-1} and its chemical properties are shown in Table 5. The rate of each nutrient to be applied as a function of expected yield and chemical soil properties are also found in the table. To determine the rate to use per yield unit (kg Mg^{-1}), Figures 10, 11, 12, 13, 14, and 15 have been used for N, P$_2$O$_5$, K$_2$O, CaO, MgO, and S, respectively.

Soil chemical property	Value	Nutrients to apply	Nutrient rates (kg ha⁻¹)
pH	6.5	$CaCO_3$	0
Mineralized N, mg kg⁻¹	30.8	N	322
Available P, mg kg⁻¹	18.4	P_2O_5	104
Exchangeable K, cmol₊ kg⁻¹	0.35	K_2O	133
Exchangeable Ca, cmol₊ kg-1	8.4	CaO	22
Exchangeable Mg, cmol₊ kg⁻¹	0.6	MgO	55
Soil exchange capacity (CEC), cmol₊ kg⁻¹	14.5	-	-
CEC used by K, %	2.4	-	-
CEC used by Ca, %	57.9	-	-
CEC used by Mg, %	4.1	-	-
Available S, mg kg⁻¹	6.2	S	38

Table 5. Soil chemical properties and nutrient rates to apply based on such properties for sowing corn where a 14 Mg ha⁻¹ yield is expected.

It can be observed in Table 5 for this example that it is liming is not necessary, and the total nutrient rates to be applied have been adjusted to both yield and soil chemical properties. In the case where residues are incorporated, their contributions must be considered as was mentioned in this chapter.

Finally, a fertilization strategy must be determined as a function of the dynamics of each nutrient in the soil-plant system, irrigation system (some nutrients can be applied via fertirrigation when there are pressurized irrigation systems), and the response of a partialized application of nutrients such as N.

3. Conclusion

Finally, the information provided in this chapter allows calculating fertilization rates in cereals using objective measurement tools, such as the productivity level and soil chemical properties, being interrelated contribute to achieving economically adequate yields with an environmental component that allows reducing the negative effects associated with incorrect nutrient use.

Author details

Juan Hirzel* and Pablo Undurraga

*Address all correspondence to: jhirzel@inia.cl

Instituto de Investigaciones Agropecuarias INIA, Centro Quilamapu, Chillan, Chile

References

[1] Laegreid, M, Bøckman, O. C, & Kaarstad, O. Agriculture, fertilizers and the environ-ment. Wallingford, UK: CABI Publishing; (1999).

[2] Hirzel, J, Matus, I, Novoa, F, & Walter, I. Effect of poultry litter on silage maize (*Zea mays* L.) production and nutrient uptake. Spanish Journal of Agricultural Research (2007). , 5, 102-109.

[3] Millner, J. P, Villaver, R, & Hardacre, A. K. The yield and nutritive value of maize hybrids grown for silage. New Zealand Journal of Agricultural Research (2005). , 48, 101-108.

[4] Herrmann, A, & Taube, F. The range of the critical nitrogen dilution curve for maize (*Zea mays* L.) can be extended until silage maturity. Agronomy Journal (2004). , 96, 1131-1138.

[5] Fernández, M. Fertilización nitrogenada y su eficiencia en maíz de grano. Simiente (1995). , 65, 122-132.

[6] Fernández, M. Influencia de la fertilización de largo plazo en el cultivo de maíz y en la residualidad de P y K en un Mollisol calcáreo. Agricultura Técnica (Chile) (1996). , 56, 107-115.

[7] Heckman, J. Sweet corn nutrient uptake and removal. HortTechnology (2007). , 17, 82-86.

[8] Hirzel, J, & Best, S. Necesidades nutricionales del cultivo de maíz en el valle regado de la VIII región. Informativo Agropecuario BIOLECHE- INIA-QUILAMAPU (2002). , 1, 5-3.

[9] Mengel, K, & Kirkby, E. Principles of plant nutrition. 4th ed. International Potash In-stitute, Worblaufen-Bern, Switzerland; (1987).

[10] Hirzel, J. Estudio comparativo entre fuentes de fertilización convencional y orgánica, cama de broiler, en el cultivo de maíz (*Zea mays* L.). Tesis Doctoral. Universidad Poli-técnica de Madrid, Madrid, España; (2007).

[11] Balmaceda, M. Fertilización cálcica en el cultivo del maíz. Tesis Ingeniero Agrónomo. Universidad de Talca; (2008).

[12] Jordan-meille, L, & Pellerin, S. Leaf area establishment of a maize (*Zea mays* L.) field crop under potassium deficiency. Plant and Soil (2004).

[13] Kim, K, Clay, D. E, Carlson, C. G, Clay, S. A, & Trooien, T. Do synergistic relation-ships between nitrogen and water influence the ability of corn to use nitrogen de-rived from fertilizer and soil? Agronomy Journal (2008). , 100, 551-556.

[14] Plénet, D, Etchebest, S, Mollier, A, & Pellerin, S. Growth analysis of maize field crops under phosphorus deficiency. Plant and Soil (2000). , 223, 117-130.

[15] Sharma, P, Chatterjee, C, Sharma, C, & Agarwala, S. Zinc deficiency and anther development in maize. Plant and Cell Physiology (1986). , 28, 11-18.

[16] Soto, P, Jahn, E, & Arredondo, S. Población y fertilización nitrogenada en un híbrido de maíz para ensilaje en el valle regado. Agricultura Técnica (Chile) (2002). , 62, 255-265.

[17] Butler, T. J, Han, K. J, Muir, J. P, Weindorf, D. C, & Lastly, L. Dairy manure compost effects on corn silage production and soil properties. Agron J. (2008). , 100, 1541-1545.

[18] Cuevas, G, Martinez, F, & Walter, I. Field-grown maize (*Zea mays* L.) with composted sewage sludge. Effects on soil and grain quality. Spanish Journal of Agricultural Research (2003). , 1, 111-119.

[19] Eghball, B, & Power, J. Phosphorus and nitrogen based manure and compost applications: maize production and soil phosphorus. Soil Science Society of America Journal (1999). , 63, 895-901.

[20] Eghball, B, Ginting, D, & Gilley, J. E. Residual effects of manure and compost applications on maize production and soil properties. Agronomy Journal (2004). , 96(2), 442-447.

[21] Hirzel, J, Walter, I, Undurraga, P, & Cartagena, M. Residual effects of poultry litter on silage maize (*Zea mays* L.) growth and soil properties derived from volcanic ash. Soil Science and Plant Nutrition (2007). , 53, 480-488.

[22] Hirzel, J, & Salazar, F. Uso de enmiendas orgánicas como fuente de fertilización en cultivos. In: Hirzel J. (ed.) Fertilización de Cultivos en Chile. Colección Libros INIA Nº28, Chillán, Chile; (2011). , 387-432.

[23] Pang, X. P, & Letey, J. Organic farming: challenge of timing nitrogen availability to crop nitrogen requirements. Soil Science Society of America Journal (2000). , 64(1), 247-253.

[24] Rees, R, & Castle, K. Nitrogen recovery in soils amended with organic manures combined with inorganic fertilisers. Agronomie (2002). , 22, 739-746.

[25] Navarro, S, & Navarro, G. Química agrícola: El suelo y los elementos químicos esenciales para la vida vegetal. Ediciones Mundi-Prensa, Madrid, España; (2003).

[26] Andraski, T, & Bundy, L. G. Corn residue and nitrogen source effects on nitrogen availability in no-till corn. Agronomy Journal (2008). , 100, 1274-1279.

[27] Coulter, J, & Nafziger, E. D. Continuous corn response to residue management and nitrogen fertilization. Agronomy Journal (2008). , 100, 1774-1780.

[28] Mellado, M. Importancia y evolución del trigo en Chile. In: Mellado M. (ed.) El trigo en Chile. Cultura, ciencia y tecnología. Colección Libros INIA N° 21. Instituto de Investigaciones Agropecuarias, Chillán, Chile; (2007). , 15-35.

[29] Benton, J. Plant nutrition manual. CRC Press LLC, Washington, USA; (1998).

[30] Hewstone, C. Producción de materia seca y absorción de macro y micronutrientes en trigo cultivado en el sur de Chile. Agricultura Técnica (Chile) (1999). , 59, 271-282.

[31] Campillo, R, Hirzel, J, & Jobet, C. Fertilización del cultivo de Trigo Panadero. In: Hirzel J. (ed.) Fertilización de Cultivos en Chile. Colección Libros INIA N°28, Chillán, Chile; (2011). , 11-80.

[32] Marschner, H. Functions of mineral nutrients: macronutrients. Part I Nutritional Physiology. In Marschner H. (ed.) Mineral nutrition of high plants. Academic Press Limited, London, England; (1990). , 195-267.

[33] Roy, R. N, Finck, A, & Blair, G. J. Tandon HIS. Plant nutrients and basics of plant nutrition. In: Roy RN. et al. (ed.) Plant nutrition for food security. A guide for integrated nutrient management. FAO Fertilizer and Plant Nutrition Bulletin 16. FAO, Rome, Italy; (2006). , 25-42.

[34] Hirzel, J. Fertilización del cultivo. In: Mellado M. (ed.) Boletín de trigo. Manejo tecnológico. Boletín INIA N° 114. Instituto de Investigaciones Agropecuarias, Chillán, Chile; (2004). , 49-75.

[35] Johnson, V, Dreier, A, & Grabouski, P. Yield and protein responses to nitrogen fertilizer of two winter wheat varieties differing in inherent protein content of their grain. Agronomy Journal (1973). , 65, 259-263.

[36] Stone, P. J, & Savin, R. Grain quality and its physiological determinants. In: Satorre EH, Slafer GA (ed.). Wheat: ecology and physiology of yield determination. Food Products Press, New York, USA; (1999). , 85-120.

[37] Haynes, R. J, & Mokolobate, M. S. Amelioration of Al toxicity and P deficiency in acid soils by additions of organic residues: a critical review of the phenomenon and the mechanisms involved. Nutrient Cycling Agroecosystem (2001). , 59, 47-63.

[38] Halvorson, A. D, Nielsen, D. C, & Reule, C. A. Nitrogen fertilization and rotation effects on no-till dryland wheat production. Agronomy Journal (2004). , 96, 1196-1201.

[39] Campillo, R, Jobet, C, & Undurraga, P. Optimización de la fertilización nitrogenada para trigo de alto potencial de rendimiento en andisoles de la Región de La Araucanía, Chile. Agricultura Técnica (Chile) (2007). , 67, 281-291.

[40] Campillo, R, Jobet, C, & Undurraga, P. Effects of nitrogen on productivity, grain quality, and optimal nitrogen rates in winter wheat cv. Kumpa-INIA in Andisols of Southern Chile. Chilean Journal of Agricultural Research (2010). , 70, 122-131.

[41] Hirzel, J, Matus, I, & Madariaga, R. Effect of split nitrogen applications on durum wheat cultivars in volcanic soil. Chilean Journal of Agricultural Research (2010). , 70(4), 590-595.

[42] Juliano, B. O. Rice in human nutrition. In: FAO (ed.) Food and Nutrition Series Nº 26. FAO, Rome, Italy. International Rice Research Institute (IRRI), Los Baños, Laguna, Philippines; (1993).

[43] Chaudary, R. C, & Tran, D. V. Speciality rices of the world: a prologue. *In* Chaudary DC, Tran DV, Duffy R. (ed.) Speciality rices of the world. Breeding, production and marketing. Food and Agriculture Organization of the United Nations. Science Publishers, Enfield, New Hampshire, USA; (2001). , 3-12.

[44] FAOEstimaciones globales de las emisiones gaseosas de NH_3, NO y N_2O provenientes de las tierras agrícolas. FAO, Roma, Italia; 2004. Available from ftp:// ftp.fao.org/docrep/fao/009/y2780s/y2780s00.pdf (accessed 10 April (2012).

[45] Alvarado, J. R. Arroz, su cultivo, situación y producción. In: Alternativas para la modernización y diversificación agrícola. Anuario del Campo. Publicaciones Lo Castillo, Sociedad Nacional de Agricultura, Asociación de Exportadores de Chile: Santiago, Chile; (1997). , 64-70.

[46] ODEPACultivos anuales: superficie, producción y rendimientos. Oficina de Estudio y Políticas Agrarias (ODEPA), Ministerio de Agricultura, Chile; 2010. http:// www.odepa.gob.cl/servlet/articulos.ServletMostrarDetalle;jsessio-nid=17656C712C5932B9501ED4DA40DD29D5?idcla=12&idn=1736.accessed 15 January (2012).

[47] Kumura, A. Studies on the effect of internal nitrogen concentration of rice plant on the constitutional factor of yield. Proceedings of the Crop Science Society of Japan (1956). , 24, 177-180.

[48] Singh, J. N, & Murayama, N. Analytical studies on the productive efficiency of nitrogen in rice. Soil Science and Plant Nutrition (1963). , 9, 25-35.

[49] Shimizu, T. Processes of yield formation in rice plants from the point of dry matter production (in Japanese). Dry Matter Production in Crops (1967). , 4, 12-26.

[50] Grist, D. H. Rice. 5th ed. Longman, London, UK; (1975).

[51] Tanaka, A, Patnaik, S, & Abichandani, C. T. Studies on the nutrition of rice plant. III. Partial efficiency of nitrogen absorbed by rice plant at different stages of growth in relation to yield of rice (*O. sativa*, var. indica). Proceedings of the Indian Academy Of Science, Section B (1959). , 49(4), 207-216.

[52] Atanasiu, N, & Samy, J. Nutrición de la planta: fertilizantes y abonos orgánicos para el arroz. In: Arroz uso eficaz de los fertilizantes. Conzett + Huber AG., Zurich, Suiza; (1985).

[53] Tinarelli, A. El arroz. Ediciones Mundi-Prensa, Madrid, España; (1989).

[54] Sheehy, J. E, Mnzava, M, Cassman, K. G, Mitchell, P. L, Pablico, P, Robles, R. P, & Ferrer, A. Uptake of nitrogen by rice studied with a ^{15}N point-placement technique. Plant and Soil (2004). , 259, 259-265.

[55] Frageria, N. K, Carvalho, G. D, & Santos, A. B. Ferreira EPB, Knupp AM. Chemistry of Lowland Rice Soils and Nutrient Availability. Communications in Soil Science and Plant Analysis (2011). , 42, 1913-1933.

[56] Trolldenier, G. Nitrogenaseaktivitaet in der Rhizosphaere von Sumpfreis in Abhaengigkeit von der Mineralstoffernaehrung. Mitteilungen der Deutschen Bodenkundlichen Gesellschaft. Goettingen (1979). , 29, 334-337.

[57] Alvarado, R, Madariaga, R, & Gómez, A. Pudrición del tallo en arroz: Respuesta varietal. IPA Quilamapu (1991). , 50, 32-35.

[58] Cordero-vásquez, A, & Murillo-vargas, J. I. Removal of nutrients by rice cultivar CR 1821 under flood irrigation. Agronomía Costarricense (1990). , 14(1), 79-83.

[59] Jokela, W. E, & Randall, G. W. Fate of fertilizer nitrogen as affected by time and rate of application on maize. Soil Science Society of America Journal (1997). , 61, 1695-1703.

[60] Jensen, L. S, Pedersen, I. S, Hansen, T. B, & Nielsen, N. E. Turnover and fate of ^{15}N-labelled cattle slurry ammonium-N applied in the autumn to winter wheat. European Journal of Agronomy (2000). , 12, 23-35.

[61] Sainz, H. R, Echeverría, H. E, & Barbieri, P. A. Nitrogen balance as affected by application time and nitrogen fertilizer rate in irrigated no-tillage maize. Agronomy Journal (2004).

[62] Sahrawat, K. Organic matter and mineralizable nitrogen relationships in wetland rice soils. Communications in Soil Science and Plant Analysis (2006). , 37, 787-796.

[63] Wienhold, B. Comparison of laboratory methods and an in situ method for estimating nitrogen mineralization in an irrigated silt-loam soil. Communication in Soil Science and Plant Analysis (2007). , 38, 1721-1732.

[64] Angus, J. F, Ohnishi, M, Horie, T, & Williams, L. A preliminary study to predict net nitrogen mineralization in a flooded rice soil using anaerobic incubation. Australian Journal of Experimental Agriculture (1994). , 34, 995-999.

[65] Wilson, C. E, Norman, R. J, & Wells, B. R. Chemical estimation of nitrogen mineralization in paddy rice soils. I. Comparison to laboratory indices. Communications in Soil Science and Plant Analysis (1994).

[66] Bushong, J. T, Norman, R. J, Ross, W. J, Slaton, N. A, Wilson, C. E, & Gbur, E. E. Evaluation of several indices of potentially mineralizable soil nitrogen. Communications in Soil Science and Plant Analysis (2007). , 38, 2799-2813.

[67] Soon, Y, Haq, A, & Arshad, M. Sensitivity of nitrogen mineralization indicators to crop and soil management. Communications in Soil Science and Plant Analysis (2007). , 38, 2029-2043.

[68] Rodriguez, A, Hoogmoed, W, & Brussaard, L. Soil quality assessment in rice production systems: establishing a minimum data set. Journal of Environment Quality (2008). , 37, 623-630.

[69] Hirzel, J, Cordero, K, Fernández, C, Acuña, J, Sandoval, M, & Zagal, E. Soil potentially mineralizable nitrogen and its relation to rice production and nitrogen needs in two paddy rice soils of Chile. Journal of Plant Nutrition (2012). , 35, 396-412.

[70] Hirzel, J, Pedreros, A, & Cordero, K. Effect of nitrogen rates and split nitrogen fertilization on grain yield and its components in flooded rice. Chilean Journal of Agricultural Research (2011). , 71, 437-444.

[71] Hirzel, J, & Cordero, K. Fertilización del cultivo de Arroz. In: Hirzel J. (ed.) Fertilización de Cultivos en Chile. Colección Libros INIA Nº28, Chillán, Chile; (2011). , 167-198.

[72] González-rojas, R. Efecto en suelos inundados de la adición de materia orgánica sobre el fósforo y otros elementos extraíbles con Mehlich III. Tesis Licenciatura en Química. Universidad de Costa Rica, Ciudad Universitaria (Costa Rica); (1997).

Chapter 7

Effects of Different Tillage Methods, Nitrogen Fertilizer and Stubble Mulching on Soil Carbon, Emission of CO$_2$, N$_2$O and Future Strategies

Sikander Khan Tanveer, Xiaoxia Wen,
Muhammad Asif and Yuncheg Liao

Additional information is available at the end of the chapter

1. Introduction

The basic reasons of greenhouse effect are the greenhouse gases which emits and absorbs the radiations. The main greenhouse gases include CO$_2$, Methane and Nitrous oxide which are playing a major role in global warming.

CO$_2$ is mainly emitted by the burning of coal, natural gas and wood etc. Before the Industrial revolution its concentration in the atmosphere was about 280 ppm but due to the burning of fossil fuels now its concentration is about 397 ppm. The concentration of CO$_2$ in the atmosphere is about 0.039 percent by volume and it is used by the plants for the process of photosynthesis.

Nitrous Oxide is also a major greenhouse gas. In the atmosphere from N$_2$O, Nitric oxide (NO) is produced, which after combination with O$_2$, reacts with Ozone. Global warming potential of this gas is 298 times more than the global warming potential of CO$_2$. In agriculture the main source of N$_2$O is the use N fertilizers, which is used for the production of crops. It is also produced from the animal wastes. In the atmosphere annually about 5.7 Tg N$_2$O-N yr^{-1}, N$_2$O is produced and agricultural soils provide about 3.5 Tg N$_2$O-N yr^{-1}, which is produced from the soils by the process of Nitrification and Dinitrification. Its emission from the soil is affected by many factors including temperature, moisture, PH, soil organic matter and soil kind etc.

The soil contains carbon in the organic and as well as in the inorganic form. Soil organic carbon is mainly present in the soil organic matter in the form of "C". and its availability in any soil mainly depends upon the soil kind, its texture, vegetation and management processes. Management of SOC is very important for the maintenance of healthy soils because its loss leads to soil infertility. SOC can be helpful in mitigating the effects of elevated CO_2 in the atmosphere, because change in land management practices can be helpful in sequestering the C from the atmosphere.

Tillage is a group of field operations and it is defined as the mechanical manipulation of the soil for improving its physical condition suitable for plant growth. The main aims of the tillage are production of a suitable tilth or soil structure, control of weeds; manage soil moisture, incorporation of organic matter, managing water and air in the soil and establishment of a surface layer which prevents the soil from wind and water erosion. A wide range of implements are used for tillage operations which vary from country to country. Different tillage implements have been designed and are used for various operations depending on the soil, type of operations, agro climatic conditions and soil conditions. Primary tillage implements include Moldboard plough, Disc plough, Chisel plough, Sub Soiler and Rotavator and secondary tillage implements include Harrows, Field cultivators and Rollers while due to recent technological developments along with economic pressures, Minimum and Zero tillage is also being practiced on large scale in the world.

Soil is the medium in which crops grow and it is one of the most precious natural resources of earth. Its maintenance for the coming generations is the responsibility of all human beings. However the urge for the production of more food, feed, fiber and fuel, especially in the form of emission of Greenhouse gases is causing irreparable damages to our environment.

2. Background

Long term records indicate increasing trends in the growth of anthropogenic greenhouse gas emissions, particularly in the last decades (IPCC, 2001). The greenhouse gases mainly include CO_2, CH_4 and N_2O having the contributions of about 60%, 20% and 6% while the potential of N_2O in warming the atmosphere is greater than 290~310 times than CO_2 and 10 times than CH_4. The concentrations of CO_2 and N_2O currently in the atmosphere are about 397 ppmv and 314 ppbv, respectively.

The emission of greenhouse gases from the soils is not clear (Le Mer J., and Roger, 2001). It has also been reported that the addition of anthropogenic greenhouse gases in to the atmosphere has been previously underestimated (Mosier A.R., et al 1998) because these gases may diffuse directly from the soil or indirectly in the atmosphere through subsurface drainage after leaching (Sawamoto T. et al. 2003).

Agricultural productivity lead to the emission of several greenhouse gases (CO_2, CH_4 and N_2O), that differ with regards on their ability to absorb the long wave radiation, and depending on their specific radiation forcing and residence time in the atmosphere. The relative ability of gases, also called global warming potential (GWP), is computed relative to carbon dioxide. The GWP is 1 for Carbon dioxide, 21 for Methane, 310 for Nitrous Oxide, 1800 for O_3 and 4000 – 6000 for CFCs (IPCC, 1995) The rate of increase in CO_2 was 1.6 ppmv per year from 1990 to 1999 (http:// www. CO_2 Science.Org/) and N_2O was 0.8 ppbv per year in the 1990s, respectively

(IPCC, 2001). According to OECD (2000), the total emission of CO_2 CH_4 and N_2O (As CO_2 equivalent) was 14142 million tons for the period from 1995 to 1997. OECD (2000) also reported that agriculture is responsible for about 1% of CO_2 emissions, about 40% of CH_4 and 60% of N_2O, so it is clear that agriculture is also playing an important role in the emission of green house gases.

Watson et al., (1995) reported that emissions of carbon dioxide (CO_2) and nitrous oxide (N_2O) are major sources of atmospheric green house gases generated from the upland agro-ecosystems. It was estimated that 90% of N_2O and 20% of CO_2 in the atmosphere come from agricultural production (Bouwman, 1990).

The global soil carbon (C) pool of 2300 Pg includes about 1550 Pg of soil Organic Carbon (SOC) and 750 Pg of soil inorganic carbon (SIC) both to 1- m depth (Batjes,1996). The soil carbon pool is three times the size of the atmospheric pool (770 Pg) and 3.8 times the size of the biotic pool (610 Pg). The SOC pool to 1-m depth ranges from 30 tons/ha in arid climates to 800 tons /ha in organic soils in cold regions and a predominant range of 50 to 150 tons/ha. The SOC pool represents a dynamic equilibrium of gains and losses. Conversion of natural to agricultural systems causes depletion of the SOC pool by as much as 60% in soils of temperate regions and 75% or more in cultivated soils of the tropics. The depletion is exacerbated when the out put of C exceeds the input and when soil degradation is severe. Some soils have lost as much as 20 to 80 tons C/ha. Severe depletion of the SOC pool degrades soil quality, reduces biomass productivity and adversely impacts water quality and the depletion may be exacerbated by projected global warming. Terrestrial ecosystems contributed to the atmospheric C enrichment during both the preindustrial and industrial areas. During the preindustrial era, the total carbon emission from terrestrial ecosystem was supposedly about twice (320 Gt or 0.04 Gt C/year for 7800 years) that of the industrial era (160 Gt or 0.8 Gt C/year for 200 years). Between 1850 and 1998, the emission from fossil fuel combustion was 270 (+ or -) 30 Gt from soil of which about one–third is attributed to soil degradation and accelerated erosion and two thirds to mineralization. The estimates of historic SOC loss range widely, from 44 to 537 Gt, with a common range of 55 to 78 Gt. The soil C can be sequestered through judicious land use and recommended management practices, which are cost effective and environmental friendly. SOC sequestration in agricultural soils and restored ecosystems depend on soil texture, profile characteristics and climate. It ranges from 0 to 150 kg C/ha per year in dry and warm regions and 100 to 1000 kg C/ha per year in humid and cool climates. Soil management techniques and land use patterns play an important role in the removal as well as store of carbon from the ecosystem and the soil management techniques are the suitable ways to reduce the CO_2 emission from the soil (IPCC, 2000; Lal, 2004). The conclusion by the intergovernmental panel on climate change (IPCC) "that there has been a discernible human influence on global climate (IPCC, 2001) is one more call for action on the reduction of green house gas (GHG) emissions. Worldwide about one fifth of the annual anthropogenic (GHG) emission comes from the agricultural sector (excluding forest conversion), producing about 5%, 70% and 50% of anthropogenic emissions of carbon dioxide (CO_2), nitrous oxide (N_2O), and 50% methane (CH_4) (Cole et al., 1996).

Nitrous Oxide (N_2O) is a natural trace gas occurring in the atmosphere that causes global warming and stratospheric ozone depletion. The concentration of atmospheric N_2O has increased up to 16% over the last 250 years at a rate of 0.25% per year (IPCC, 2007) and agricultural soils account for approximately 42% of anthropogenic N_2O emissions (IPCC,2007) and nitrogen fertilization is considered as a primary source of N_2O emissions from agricultural soils (Mosier et al., 1998; Mosier and Kroeze,2000). The annual global emission of N_2O from soils is estimated to be 10.2 Tg N or about 58% of all emissions (Mosier A.R., et al 1998).

Use of nitrogen fertilizers increased at the rate of 6-7% per year during the 1990s (Mosier, 2004). According to FAO (2008) during 2008 to 2012 on world level the demand for nitrogen fertilizers on annual basis will increase at the rate of 1.4% and about 69% of this growth will take place in Asia. It has been reported by many researchers (e.g. Hatano R., and Sawamoto T, (1991), Kaiser E, A., and Ruser R, (2000), and Mosier A.R., and Delgado J.A., (1997) that N_2O emission increases with increasing nitrogen fertilization. Nitrogen fertilizer is one most important mineral fertilizer, both in the amount of plant nutrient used in agriculture and in energy requirements. Its principal products include Ammonia, Urea, Ammonium Nitrate, Urea/ Ammonium nitrate solution, Di-Ammonium Phosphate and Ammonium Sulphate. Nitrogen fertilizers are energy–intensive. For example one kilogram of nutrient-N requires about 77.5 MJ for its manufacture, packing, transportation, distribution, and application (Stout, 1990). For its production manufacturer requires pure gaseous nitrogen and hydrogen. AS compared to hydrogen, pure gaseous nitrogen is simple and inexpensive. Natural gas and coal are the main sources of hydrogen for fertilizer production (Helsel, 1992). In addition to this, in some developing countries transportation routes can be very energy demanding.

N_2O emission is affected by many factors but the most important of these are tillage and fertilization. Increased N_2O emission from No-tilled soils as compared to the tilled soils have been reported by many researchers (Aulakh M.S. etal,1984, Jacinthe P.A., and Dick W.A., 1997, Lal R. et al. 1995, Mackenzie A.F., and Fan M.X., Cadrin F., 1997 and Mummey D.L.,etal, 1997). However Grandy et al. (2006) reported that N_2O emissions were similar between NT and CT systems. Lemake et al. (1999) observed lower N_2O emissions from NT compared to CT in some north-central Alberta soils. Similarly Hao et al. (2001) observed decline in N_2O emissions when crop residue was retained, especially in plots tilled in autumn after crop harvest. Mc Swiney and Robertson (2005) and Wanger Riddle et al. (2007) have reported that fertilizer application rates influence the N_2O fluxes. They also reported that the best management practices can be helpful in reducing the emission of N_2O.

Nitrous oxide (N_2O) is a major green house gas (GHG), although its emission is numerically smaller as compared with the other greenhouse gases (Isermann,1994), but global warming potential (GWP) of it is about 296 times higher than that from Carbon dioxide (CO_2). It is reported that arable soils are responsible for about 57% of the annual N_2O emissions in the world (Moiser et al.,1998).Effects of seasons (Van Kessel et al. 1993; Nyborg et al.1997 and Aulakh et al.1982), nitrogen fertilizers (Mc Kenney etal.1980; Eichner 1990), manure or legumes (Aulakh etal.1991; Laidlaw 1993) on emission of N_2O have also been reported.

Low fertilizer use efficiency is mainly due to the inexpert use of fertilizer and it is associated with the water and air pollution. These kinds of symptoms are common in many countries, where due to leaching and volatilization important fertilizer losses are taking place (FAO, 2000). In such regions the extra use of fertilizers can be reduced by adopting the more efficient methods of fertilizers applications, use of fertilizers at the proper timings to meet the nutrient demand of the crop, nitrification inhibitors, controlled release fertilizers, optimization of till-age, irrigation and drainage and ultimately the use of precision farming can be helpful in saving the extra use of fertilizers (Viek and Fillery, 1984; IPCC, 1996). Higher productivity can be achieved at a given level of fertilizer use with improved crop and fertilizers management. However, there are many constraints including economic, educational and social in improving the fertilizer productivity.

Conventional tillage increases the CO_2 in the atmosphere by promoting the loss of soil organic matter, while instead of conventional tillage, conservation tillage increases soil organic matter (SOM) with the passage of time (Dao, 1998) and available water content and soil aggregation (Pare etal.,1999). A lot number of researchers (La Scala Jr et al (2001), Lipiec J., and Hatano R., (2004), Reicosky D.C et al (1997), and Sanchez M.L et al (2002)) have reported that emission of CO_2 is lower in no-till or reduced tillage as compared to the conventional tillage. On the other side more carbon sequestration is in the soil which is under no-till or reduced tillage as compared to the soil under conventional tillage (Ball B.C et al (1999) and McConkey Liang B.C et al (2002)). Wilson H.M and M.M. Al-Kaisi (2010) reported 23% less emission of CO_2 from the soil fertilized with 270 kg N/ha as compared to the soils fertilized with 0 and 135 kg N/ha in a continuous corn and a corn-soybean rotation. Similarly Burton et al. (2004) found that N fertilized plots averaged 15% less soil CO_2 emissions than unfertilized plots.

Campbell et al. (2001), reported that without adequate fertility from Conservation to No-till-age may not always result in an increase in soil N or C. The contribution of CO_2 in the total atmospheric greenhouse effect is about 60% (Rastogi M., et al 2002). Conversions of native soils for agricultural use have contributed a large to the emission of CO_2 in to the atmosphere (Paustian K., et al, 1998).

Agriculture seems to have potential to make an important contribution to the mitigation of global climate change. Lal et al. (1998) estimated that changes in global agricultural practices can be able to sequester over about 200 million metric tons of carbon (Mt C) per year. Changes in agronomic practices in United States are thought to have the potential to offset nearly 10% of its total carbon emissions (FAO, 2001).The International Panel on Climate Change (IPCC, 2000) quotes figures showing that alone conservation tillage can be able to store more than a ton of carbon per hectare per year, while other researchers have provided the figures that range from a low of 3 to a high of 500 kg C per hectare per year (Uri, 2001; Follet, 2001).

Uri, (2001) and West and Marland, (2002), reported that No-till cultivation that appears to bring about carbon benefits as compared to the different kinds of tillage systems, but it increases cost of production (because more chemical inputs are required) and often reduces yields (Lerohl and Van Kooten,1995).

It is generally acknowledged that soil carbon will be increased by the adapting of No (Zero) tillage as compared to the Conventional (Intensive) tillage practices (Kern and Johnson, 1993; IPCC, 2000; Uri, 2001). The relationship between NT and carbon storage is complex one. A lot number of researchers have examined the effects of crop type, fertilizers and rotation (Campbell et al., 2001), climate and soil texture (Tobert et al., 1998; Six et al., 1999) and time (Ding et al., 2002) on carbon storage potential.

The way by which Conventional Tillage (CT) might store more carbon than No Tillage (NT) is not clear (Angers et al., 1997). Conventional tillage increases CO_2 respiration as the soil is plowed (Lupwayi et al., 1999), but it appears that due to plowing organic matter is pushed more deeply in to the soil, which in future facilitates the adsorption and stabilization of more organic material into the soil, as compared to the way in which straw and residue remain concentrated on top of the ground (Paustian et al., 1997).

Tillage often decreases soil organic matter (Gebhart et al., 1994) and it increases the flux of CO_2 from the soils (Reicosky and Lindstrom, 1993) through enhanced biological oxidation of soil carbon by increasing subsequent microbial activity as a result of residue incorporation (Reicosky et al., 1995).

In the tropics the conversion of native ecosystems to agricultural use is believed to be the largest non- fossil fuel source of CO_2 input to the atmosphere. The production of more fertilizers increases the CO_2 emission, but its use reduces the need of further expansion into forested areas and may allow land to set aside for revegetation or reforestation. In addition to it modern cultivation practices and injudicious use of mineral fertilizers are further deteriorating our soil fertility. The intensive management of agricultural soils has resulted in the depletion of soil carbon (C) stocks and has increased atmospheric carbon dioxide (CO_2) levels. Lal and Bruce (1999) and Lal, 2003, 2004 reported that conservation tillage can increase the amount of C sequestered in agricultural soils. A review of soil organic carbon (SOC) studies from West and Post (2002) concluded that on an average conservation tillage could sequester 0.60+ 0.14 t C per ha per year. It has also been reported by several recent published studies that conventional or reduced tillage systems have little to no difference in soil organic carbon (SOC), (Dolan et al.,2006; Venterea et al; 2006; Baker et al., 2007; Blanco- Canqui and Lal, 2008).

A number of studies have reported that conservation tillage systems have higher N_2O emissions when compared to the conventional tilled systems (Robertson et al., 2000; Mummey et al., 1998; Ball et al., 1999).

The use of organic waste can be helpful in improving the crop productivity, improving the soil health, reduce the waste disposal problem and the betterment of environment. In some areas of the world wheat and maize straw are produced in such a huge quantity that their disposal is a big problem. Though a large portion of them is used as animal feed or for making compost but a large portion of it is being burnt which is a big threat to the environment. Therefore possible use of crop residues into the farm soil must be explored.

Researchers have also debated about the burning of crop residues as compared to their burning on the carbon flux. Clapp et al. (2000) and Duiker and Lal (2000) favor the leaving the straw, where as Sanford et al. (1982) found that straw limits the yields. Dalal (1989) reported that burning of residues contribute to carbon sequestration at depths as low as 0.9- 1.2 m.

Javed et al. (2009), reported that N addition with straw increases the CO_2 flux and sequestrates the soil C by mitigating the global carbon budget. Since SOC storage is highly dynamic in gain and loss of soil organic matter (SOM), manure abandonment and straw removal led to loss of soil organic carbon (SOC) (Qiu et al., 2004; Tang et al., 2006). Field practices with low carbon inputs to arable soils, removal of crop straw and manure abandonment have depleted SOC contents (Wang et al., 2008). The loss of SOC has bad affects on biological, physical and chemical properties (Kumar and Goh 2003). Returning crop straw to the soil can enhance the crop yield by affecting the microbial processes and nutrient availability (Olivier et al., 2000). Increase of SOC storage in crop land improves soil productivity and is good for healthy environment (Lal, 2004). Sequestration of atmospheric CO$_2$ will be significantly improved, if large quantities of crop residues and organic manures are returned to the soil (Lal, 2004 & Lal, 2002). Increase in SOC storage in crop land improves soil productivity and improves the environmental health. Thus can be recognized as win-win strategy (Lal, 2004).

Increase in soil organic matter works as a sink for atmospheric CO$_2$ and thus reduces the adverse effects of global warming (Lal, 2004). Addition of straw evolutes CO$_2$, affects microbial activity for decomposition, recycles available nutrients and restores soil organic carbon (SOC) (Henriksen and Breland, 1999; Hadas et al., 2004).

Wu et al. (2003). reported that 57% of cultivated soils in China have experienced losses in soil organic carbon (SOC) due to the injudicious use chemical fertilizers and intensive cultivation along with less use of organic manures and crop straw since the introduction of synthetic fertilizers in 1950s.

The burning or discarding of the crops straw has caused a decline in soil organic matter (SOM), reduction in microbial activity and is causing pollution by discharging CO$_2$ into the atmosphere. Michellon and Parret, (1994) reported that the use of crop residue can be helpful in the reduction of chemical fertilizer use. While Boyer et al., (1996) reported that the use of crop residues can be helpful in the restoration of soil fertility. Lal et al. (1998) recommended the rate of residue for cool and humid areas about 400-800 kg/ha/year, while about 200-400 kg for warm and dry areas.

Kumar and Goh, 2000 reviewed the effects of crop residues and management practices on the soil quality, soil nitrogen dynamics recovery and as well as on crop yields. They reported that residues of cultivated crops are a significant factor for crop production through their effects on soil health as well as on soil and water quality. Unger et al. (1988), reviewed the role of surface residues on water conservation and they reported that surface residues enhance water infiltrations. Dao, (1993) and Hatfield & Pruger, (1996), reported that surface mulch helps in reducing water losses from the soil and promotes biological activity which enhances nitrogen mineralization especially in the surface layers. Leak (2003) reported that rotations increase microbial diversity. Jacinthe P.A et al. (2002) reported that carbon sequestration can be enhanced by the crop residue management.

3. Future strategies

Many crop management and soil management practices in the future can be helpful in the reduction of emission of greenhouse gases along with the maintenance of soil fertility, few of them include

1. Adaptation of tillage systems that are helpful in the reduction of emission of greenhouse gases, and can be helpful in the increasing the soil fertility.

2. Selection of more better cropping systems which are more environmental friendly and give maximum crops productions.

3. Judicious and timely use of nitrogenous fertilizers

4. Use of crop residues and animal manures

5. Use of crop rotations

6. Reduce the soil erosion

In future from the less available land resources, only the better crop production practices will be helpful to feed the burgeoning population. Instead of traditional tillage practices, use of productive but more sustainable environmental friendly management practices will resolve this problem. Crop and soil management practices that maintain soil health and reduce farmers cost are essential in this regard. Minimum soil disturbance, soil cover (mulch) combined with judicious use of fertilizers will be helpful in getting the maximum yields of crops and maintenance of healthy environment.

Author details

Sikander Khan Tanveer[1], Xiaoxia Wen[1], Muhammad Asif[2] and Yuncheg Liao[1*]

*Address all correspondence to: yunchengliao@163.com or yunchengliao@nwsuaf.edu.cn

1 College of Agronomy, Northwest A&F University Yangling, Shaanxi, China

2 Agriculture, Food and Nutritional Science, 4-10 Agriculture/Forestry Centre, Univ. of Alberta, Edmonton, Canada

References

[1] Angers, D. A, Bolinder, M. A, Carter, M. R, Gregorich, E. G, Drury, C. F, Liang, B. C, Wrony, R. P, Simard, R. R, Donald, R. G, Beyaert, R. P, & Martel, J. (1997). Impact of tillage practices on organic carbon and nitrogen storage in cool, humid soils of eastern Canada. Soil Tillage Res. , 41, 191-201.

[2] Arah JRMSmith, K.A, Crichton I.J,LI, HS.: (1991). Nitrous oxide production and denitrification in Scottish arable soils. European Journal of Soil Science, , 42, 351-367.

[3] Aulakh, M. S, Rennie, D. A, & Paul, E. A. (1984). Gaseous nitrogen losses from soils under zero-till as compared with conventional-till management systems J. Environ. Qual, , 13, 130-136.

[4] AulakhM.S; Doran, J. W., Walters, D.T., and Power, J.F.(1991). Legume residue and soil water effects on denitrification in soils of different textures. Soil Biol. Biochem. , 23, 1161-1167.

[5] AulakhM.S; Rennie D.A.and Paul,E.A.(1982). Gaseous nitrogen losses from cropped and summer-fallowed soils. Can. J. Soil Sci., 62, 187-196.

[6] Baker, J. M, Ochsner, T. E, Venterea, R. T, & Griffis, T. J. (2007). Tillage and soil carbon sequestration: what do we really know? Agriculture, Ecosystems and Environment., 118, 1-5.

[7] Ball, B. C, Scott, A, & Parker, J. P. O, CO_2 and CH_4 fluxes in relation to tillage, compaction, and soil quality in Scott land. Soil and Tillage Res. , 53, 29-39.

[8] Batjes, N. H. and N in soils of the world. European Journal Soil Science. , 47(2), 151-163.

[9] Bayoumi HamudaH.E.A.F., Kecskes, M.(2003). Correlation between the efficiencies of CO_2 release, FDA, and dehydrogenase activity in the determination of the biological activity in soil amended with sewage sludge. Nyiregyhaza, Hungary. , 11-16.

[10] Bhattaacharyya, R, Kundu, S, Prakash, R, & Gupta, H. S. (2008). Sustainability under combined application of minerals and organic fertilizers in a rainfed soybean-wheat system of the Indian Himalayas. European Journal of Agronomy, , 28, 33-46.

[11] Blanco-canqui, H, & Lal, R. (2008). No tillage and soil profile carbon sequestration: An on- farm assessment. Soil Sci. Soc. Am. J. , 72, 693-701.

[12] Bouwman, A. F. (1990). Exchange of greenhouse gases between terrestrial ecosystems and the atmosphere, soils and the Greenhouse Effect. John Wiley and Sons, Chichester, England, , 61-127.

[13] Boyer, J, Michellon, R, & Lavelle, P. (1996). Characterization of soil Pelargonium asperum with different management options. In: Proceedings of the XII th International Colloquium on soil Zoology. Dublin. , 236.

[14] Burton, A. J, Pregitzer, K. S, Crawford, J. N, Zogg, G. P, & Zak, D. R. (2004). Simulated chronic deposition reduces soil respiration in Northern hardwood forests. Glob. Ch. Biol., 10(3-N), 1080-1091.

[15] Campbell, C. A, Selles, F, Lafond, G. P, & Zentner, R. P. (2001). Adopting zero tillage management impact on soil C and N under long-term crop rotations in a thin Black Chernozem. Can. J. Soil Sci. , 81, 139-148.

[16] Campbell, C. A, Selles, F, Lafond, G. P, & Zentner, R. P. (2001). Adopting zero tillage management: Impact on soil C and N under long-term crop rotation under different stubble and tillage practices. Australian J. Soil Res. , 30, 71-83.

[17] Clapp, C. E, Allmaras, R. R, Layese, M. F, Linden, D. R, & Dowdy, R. H. (2000). Soil organic carbon and 13 C abundance as related to tillage, crop residue, and nitrogen fertilization under continuous corn management in Minnesota. Soil Tillage Research. 53 (3/4), 127-142.

[18] Cole, C. V, Duxbury, J, Freny, J, Heinemeyer, O, Mianmi, K, Rosenberg, N, Sampson, N, Saurbeck, D, & Zhao, Q. (1997). Global estimates of potential mitigation of greenhouse gas emissions by agriculture. Nutrient Cycling in Agro ecosystems , 49, 221-228.

[19] Cole, V, Cerri, C, Minami, K, Moiser, A, Rosenberg, N, & Sauerbeck, D. (1996). Agricultural options for mitigation of greenhouse gas emissions, in R.Watson, M. Zinyowera and R. Moss (eds.), "Climate Change 1995: Impacts, Adaptations and Mitigation of Climate Change: Scientific-Technical Advances", Contribution of Working Group II to the second Assessment of the IPCC, Cambridge, Cambridge University Press.

[20] Dalal, R. C, Wang, W. J, Roertson, G. P, & Parton, W. J. (2003). Nitrous oxide emission from Australian agricultural lands and mitigation options: a review. Australian Journal of Soil Research, , 41, 165-195.

[21] Dalal, R. C. (1989). Long-term effects of no-tillage, crop residue, and nitrogen application on properties of vertisol. Soil Sci. Soc. Amer. J. , 53, 1511-1515.

[22] Dao, T. H. (1993). Tillage and winter wheat residue management effects on water infiltration and storage. Soil Science Society of America Journal., 57, 1586-1595.

[23] Dao, T. H. (1998). Tillage and crop residue effects on carbon dioxide evolution and carbon storage in a Palenstoll.Soil Sci. Soc. Am.J., 62, 250-256.

[24] Ding, G, Novak, J. M, Amarasiriwardena, D, Hunt, P. G, & Xing, B. (2002). Soil organic matter characteristics as affected by tillage management. Soil Sci. Soc. Amer, J., 66, 421-429.

[25] Dolan, M. S, Clapp, C. E, Allmaras, R. R, Baker, J. M, & Molina, J. A. E. (2006). Soil organic carbon and nitrogen in a Minnesota soil as related to tillage, residue and nitrogen management. Soil tillage. Research. , 89, 221-231.

[26] Duiker, S. W, & Lal, R. (2000). Carbon budget study using CO_2 flux measurements from a no till system in central Ohio. Soil Tillage. Research.54 (1/2), 21-30.

[27] Eichner, M. J. (1990). Nitrous oxide emissions from fertilized soils: Summery of available data. J. Environ. Qual. , 19, 272-280.

[28] FAO: "Food and fuel in a warmer World" News & HighlightsFAO Newsletter, December 4, (2001).

[29] FAO: Current World Fertilizer Trends and Outlook to (2004). Rome.

[30] Farquhar, G. D, Ehleringer, J. R, & Hubickn, K. T. (1989). Carbon isotope discrimination and photosynthesis. Annu. Rev. Plant Physiol. Plant Mol. Biol. , 40, 503-537.

[31] Follett, R. F. (2001). Soil management concepts and carbon sequestration in cropland soils. Soil Tillage Research.61(1/2), 77-92.

[32] Food and Agricultural Organization of the United Nations(2008). FAOSTAT Database http://www. Fao.org/ faostat/.

[33] Gebhart, D. L, Johson, H. B, Mayeux, H. S, & Polley, H. W. (1994). The CRP increases soil organic carbon. J. Soil water conserv., , 49, 488-492.

[34] Grandy, A. S, Loecke, T. D, Parr, S, & Robertson, G. P. (2006). Long term trends in nitrous oxide emissions, soil nitrogen, and crop yields of till and no-till cropping systems. J. Environ. Qual., 35, 1487-1495.

[35] Grattan, S. R, Berenguer, M. J, Connell, J. H, Polito, V. S, & Vossen, P. M. (2006). Olive oil production as influenced by different quantities of applied water. Agric.Water Manage. , 85, 133-140.

[36] Hadas, A, Larissa, K, Mustafa, G, & Emine, E K. (2004). Rates of decomposition of plant residues and available nitrogen in soil, related to residue composition through simulation of carbon and nitrogen turnover. Soil Biology and Biochemistry , 36, 255-266.

[37] Hao, X, Chang, C, Carefoot, J. M, & Ellert, H. H. B.H., (2001). Nitrous oxide emissions from an irrigated soil as affected by fertilizer and straw management. Nutr. Cycl. Agroecosyst., 60, 1-8.

[38] Hata Hatano RSawamoto T.: (1997). Emission of N_2O from a clayey aquic soil cultivated with onion plants. In: Ando, T. et al (Eds), Plant Nutrition- for sustainable Food Production and Environment, , 555-556.

[39] Hatfield, K. L, & Prunger, J. H. (1996). Microclimate effects of crop residues on biological process. Theor.Appl.Climatol., 54, 47-59.

[40] Helsel, Z. R. (1992). Energy and alternatives for Fertilizer and Pesticide use, in R.C. Fluck (ed.), Energy in World Agriculture 6. Amesterdam, Elsevier, , 177-201.

[41] Henriksen and Breland(2002). Carbon mineralization, fungal and bacterial growth and enzyme activities as affected by contact between crop residues and soil. Biol Fert Soil., 35, 41-48.

[42] Huang, Y, Zou, J, Zheng, X, Wang, Y, & Xu, X. (2004). Nitrous oxide emissions as influenced by amendment of plant residues with different C: N ratios. Soil Biology and Biochemistry., 36, 973-981.

[43] Intergovernmental panel on Climate Change (IPCC)(2007). Chapter 8: Agriculture. In: Metz, B., Davidson, O., Bosch, P., Dave, R., Meyer, L. (Eds.), Climate Change 2007: Mitigation.Contribution of Working Group III to the Fourth Assessment Report of the Intergovernmental Panel on Climate Change. Cambridge University Press, Cambridge, United Kingdom and New York.

[44] Intergovernmental panel on Climate Change (IPCC): Summery for PolicymakersA report of working group I of the Intergovernmental Panel on Climate Change, 1-83(http://WWW.ipcc.ch/pub/pub.htm), (2001).

[45] IPCC ((2000). Special report on emissions scenariosCambridge University Press, Cambridge, UK.

[46] IPCC (2007) Climate Change (2007). The Physical Science BasisCambridge University Press, Cambridge, UK and New York, USA.

[47] IPCC (Intergovernmental Panel on Climate Change):(2000). Land use, Land Use Change, and Forestry. Cambridge University Press, New York. Kern, J. and Johnson, M.: 1993. Conservation tillage impacts on national soil and atmospheric carbon levels. Soil Sci. Soc. Amer. J. , 57, 200-210.

[48] IPCC(2001). Atmospheric chemistry and greenhouse gases. Climate Change 2001: The Scientific Basis, Hougton et al. Eds., Cambridge University Press, Cambridge, , 248-253.

[49] IPCC: (1995). Climate Change 1995.Working group1.IPCC, Cambridge University Press, Cambridge, U.K.

[50] IPCC: (1996). Technologies, Policies and measures for Mitigation Climate Change, Cambridge, Cambridge University Press.

[51] IPCC: 2001Climate Change (2001). Impacts, adaptations, and Vulnerability. Contribution of Working group II to the Third Assessment Report of the Intergovernmental Panel on Climate Change (IPCC), Cambridge.Cambridge University Press, 1000 pp.

[52] Iqbal, J, Hu, R, Lin, S, Bocar, A, & Feng, M. (2009). Carbon dioxide emission from Ultisol under different land uses in mid- subtropical China. Geoderma , 152, 63-73.

[53] Isermann, K. (1994). Agricultures share in the emission of trace gases affecting the climate and some cause- oriented proposals for sufficiently reducing this share.Environ.Pollut., 83, 95-11.

[54] Jacinthe, P. A, & Dick, W. A. (1997). Soil management and nitrous oxide emissions from cultivated fields in southern Ohio. Soil Tillage Research, , 41, 221-235.

[55] Jacinthe, P. A, Lal, R, & Kimble, J. M. Carbon budget and seasonal carbon dioxide emission from a central Ohio Luvisol as influenced by wheat residue amendment. Soil Tillage Research.(2002). , 67, 147-157.

[56] Janzen, H. H, Campbell, C. A, Izaurralde, R. C, Ellert, B. H, Juma, N, Mcgill, W. B, & Zentener, R. P. (1998 b). Management effects on soil C storage in the Canadian prairies. Soil Tillage Research. (In press).

[57] Janzen, H. H, Campbell, C. A, Izaurralde, R. C, & Ellert, B. H. (1998a). Soil carbon dynamics in Canadian Argo systems. in R. Lal, J.M. Kimble, R.S. Follett, and B.A. Stewart, eds. Soil processes and carbon cycle. CRC Press, Boca Raton, FL., 57-80.

[58] Kaiser, E, & Ruser, A. R.: (2000). Nitrous oxide emissions from arable soils in Germany- An evaluation of six long-term field experiments. J. Plant Nutr. Soil Sci., , 163, 249-260.

[59] Kern and Johnson:(1993). Conservation tillage impact on national soil and atmospheric carbon levels. Soil Sci.Soc. Amer. J. , 57, 200-210.

[60] Kumar and Goh: (2003). Nitrogen release from crop residues and organic amendments as affected by biochemical composition. Soil Sci. plant Anna , 34, 2441-2460.

[61] Kumar K & Goh KM(2000). Crop residues and management practices: effects on soil, soil nitrogen dynamics, crop yield and nitrogen recovery. Advances in Agronomy 6:

[62] La Scala JrN., Lopes A., Marques Jr., Pereira G.T.: (2001). Carbon dioxide emissions after application of tillage systems for a dark red latosol in southern Brazil. Soil Tillage Research, , 62, 163-166.

[63] Laidlaw, J. W. (1993). Denitrification and nitrous oxide emission in thawing soil. M.Sc. Thesis, Department of Soil Science, University of Alberta, Admoton, AB.

[64] Lal, R, Fausey, N. R, & Eckert, D. J. Land use and soil management effects on emissions of radiatively active gases in two Ohio Soils. In: Lal, R., Kimble, J., Levine, E., Stewart, B. (Eds.), Soil Management and Greenhouse Effect. Lewis/CRC Publ., Boca Raton, FL, (1995). , 41-59.

[65] Lal, R. (2003). Global potential of soil carbon sequestration to mitigate the green house effect. Crit. Rev. Plant Sci., 22, 151-184.

[66] Lal, R. (2004). Soil carbon sequestration to mitigate climate change, Geoderma , 123, 1-22.

[67] Lal, R, & Bruce, J. P. (1999). The potential of world cropland soils to sequester C and mitigate the green house effect. Environ. Sci. Pollut., 2, 177-185.

[68] Lal, R, Kimble, L. M, Follett, R. F, & Cole, C. V. (1998). The potential of U.S Cropland to Sequester C and Mitigate the Greenhouse Effect, Ann Arbor Press, Chelsea, MI.

[69] Lal, R. (2004). Is crop residue a waste? J. Soil water Conservation. , 59, 136-139.

[70] Lal, R. (2004). Soil carbon sequestration impacts on global climate change and food security. Science , 304, 1634-1627.

[71] Le Mer JRoger P.: Production, oxidation, emission and consumption of methane by soils: A review. Eur. J. Soil. Biol, (2001). , 37, 25-50.

[72] Leake, A. R. (2003). Integrated pest management for conservation agriculture. In. Garc Benites J, Martinez-Vilela A. (editors).Conservation Agriculture, Environment, Farmers experiences, Innovations,Socio-economy, Policy. Kluwer. A publishers. Dordrecht/Boston/London., 271-279.

[73] Lemke, R. L, Izaurralde, R. C, Nyborg, M, & Solberg, E. D. (1999). Tillage and N source influence soil-emitted nitrous oxide in the Alberta Parkland region. Canadian Journal of Soil Science, , 79, 15-24.

[74] Lerohl, M. L, & Van Kooten, G. C. (1995). Is soil erosion a problem on the Canadian Prairies? Prairie Forum , 20, 107-121.

[75] Li, C. S, Frolking, S, & Butterbach-bahl, K. (2005). Carbon sequestration in arable soils is likely to increase nitrous oxide emissions, offsetting reductions in climate radiative forcing. Climate Change, , 72, 321-338.

[76] Linn, D. M. And Doran, J.W.(1984). Effect of water- filled pore space on carbon dioxide and nitrous oxide production in tilled and non-tilled soils. Soil Sci. Soc. Am. J. , 48, 1267-1272.

[77] Lipiec, J, & Hatano, R. Effects of soil compaction on greenhouse gas fluxes. In: Glinski J., Jozefaciuk G., Stahr K. (Eds.) Soil-Plant-Atmosphere: Aeration and Environmental problems, (2004). , 18-29.

[78] Luwayi, N. Z, Rice, W. A, & Glayton, G. W. (1999). Soil microbial biomass and carbon dioxide flux under wheat as influenced by tillage and crop rotation", Can.J.Soil Sci, 19(2)., 273-280.

[79] Mackenzie, A. F, Fan, M. X, & Cadrin, F. Nitrous oxide emission as affected by tillage, corn-soybean-alfalfa rotations and nitrogen fertilization. Can. J. Soil Sci., (1997). , 77, 145-152.

[80] Maljanen, M, Martikainen, P. J, Aatlonen, H, & Silova, J. (2002). Short term variation in fluxes of carbon dioxide and methane in cultivated and forested organic boreal soils. Soil Biology& Biochemistry, , 34, 577-584.

[81] McConkey Liang BC., Campbell C.A., Curtin D., Moulin A., Brandt S.A., Lafond G.P.: Crop rotation and tillage impact on carbon sequestration in Canadian prairie soils. Soil Tillage Research, (2003). , 74, 81-90.

[82] Mckenny, D. J, Shuttleworth, K. F, & Findlay, W. I. (1980). Nitrous oxide evolution rates from fertilized soil: Effects of applied nitrogen. Can. J. Soil Sci., 60, 429-438.

[83] Mcswiney, C. P, & Robertson, G. P. (2005). Non linear response of N_2O flux to incremental fertilizer addition in a continuous maize (Zea mays L.) cropping system. Global Change Biol. , 11, 1712-1719.

[84] Michellon, R, & Perret, S. (1994). Conception de systems agricoles durables avec couverture herbacee permanente pour les haunds de la Reunion. Centre de cooperation internationale en recherché agronomique pour le development- reunion, Montpillier, France.

[85] Moiser, A. R, Delgado, J. A, & Keller, M. (1998). Methane and nitrous oxide fluxes in an acid Oxisol in Western Puerto Rico: Effects of tillage, liming and fertilization. Soil Biol. Biochem. , 30, 2087-2098.

[86] Moiser, A. R, & Kroeze, C. (2000). Potential impact on the global atmospheric N_2O budget of the increased nitrogen input required to meet future global food demands, Chemposh. Global Change Sci., 2, 465-473.

[87] Moiser, A. R, Kroeze, C, Nevison, C, Oenema, O, Seitzinger, S, & Van Cleemput, O. (1998). Closing the global N_2O budget: Nitrous oxide emissions through the agricultural nitrogen cycle. Nutr. Cycl. Agroecosyst., , 225-248.

[88] Moiser, A. R. (2004). Agriculture and the Nitrogen Cycle. Assessing the impacts of fertilizer use on feed production and the environment. SCOPE Series Island Press, Washington D C.(65)

[89] Mosier, A. R, Kroeze, C, Nevison, C, Oenema, O, Seitzinger, S, & Van Cleemput, O. (1998). Closing the global N2O budget: nitrous oxide emissions through the agricultural nitrogen cycle. Nutr. Cycling Agroecosyst., , 52, 225-248.

[90] Mummey, D. L, Smith, J. L, & Bluhm, G. (1998). Assessment of alternative soil management practices on N_2O emissions from US agriculture. Agri. Ecosystem. Environ. , 70, 79-87.

[91] National Bureau of StatisticsChina, (2008). China Statistical Yearbook- 2008. China Statistics Press, Beijing (In Chinese)

[92] Nyborg, M, Laidlaw, J. W, & Solberg, E. D. And Malhi, S.S. (1997). Denitrification and nitrous oxide emissions from soil during spring thaw in a Malmo Loam, Alberta. Can. J. Soil Sci. , 77, 153-160.

[93] OECD(2000). Environmental indicators for agriculture- methods and results. Exactive summery 2000, Paris.

[94] Oliver, C D, & William, R H. (2000). Decomposition of rice straw and microbial carbon use efficiency under different soil temperature and moistures. Soil Biology and Biochemistry, , 32, 1773-1785.

[95] Pare, T, Dinel, H, Moulin, A. P, & Townley- Smith, L. Organic matter quality and structural stability of a black Cherno zemic soil under different manure and tillage practices. Geoderma , 91, 311-326.

[96] Paustian, K, Andren, O, Janzen, H. H, et al. (1997). Agricultural soils as a sink to mitigate CO_2 emissions. Soil use and Management, , 13, 230-244.

[97] Paustian, K, Amdren, O, Janzen, H. H, Lal, R, Smith, R, Tian, P, Tiessen, G, Van Noordwijk, H, & Woomer, M. P.L.: (1997). Agricultural soils as a sink to mitigate CO_2 emissions. Soil Use Mgmt.134 (4, supp.), , 230-244.

[98] Paustian, K, Elliott, E. T, & Carter, M. R. Tillage and crop management impacts on soil C storage: Use of long-term experiment data. Soil Tillage Research, (1998). , 47, 7-12.

[99] Powlson, D. S, Witmore, A. P, & Goulding, W. T. (2011). Soil carbon sequestration to mitigate climate change: a critical re-examination to identify the true and the false. European Journal of Soil Science, , 62, 42-45.

[100] Rastogi, M, Singh, S, & Pathak, H. Emission of carbon dioxide from soil. Current Sci., (2002). , 82, 510-517.

[101] Reicosky, D. C, Dugas, W. A, & Torbert, H. A. (1997). Tillage-induced soil carbon dioxide loss from different cropping systems. Soil Tillage Research, , 41, 105-118.

[102] Reicosky, D. C, Kemper, W. D, & Langdale, G. W. Douglas, Jr., C.L. and Rasmussen, P.E., (1995). Soil organic matter changes resulting from tillage and biomass production. J. Soil Water Conserv., , 50(3), 253-261.

[103] Reicosky, D. C, & Lindstorm, M. J. (1993). Fall tillage method: Effect on short-term carbon dioxide flux from soil. Agron. J., , 85(6), 1237-1243.

[104] Robertson, G. P, Paul, E. A, & Harwood, R. R. (2000). Greenhouse gases in intensive agriculture: contributions of individual gases to the radioactive forcing of the atmosphere. Science , 289, 1922-1925.

[105] Sanchez, M. L, Ozores, M. I, Colle, R, Lopez, M. J, De Torre, B, Garcia, M. A, & Perez, I. fluxes in cereal land use of the Spanish plateau: influence of conventional and reduced tillage practices. Chemosphere, , 47, 837-844.

[106] Sanford, J. O, Hariston, J. E, & Reinschmiedt, L. L. (1982). Soybean-Wheat Double Cropping: Tillage and Straw Management Glycine max, Triticum aestivum, Yield, (osl), Returns, Mississippi", Research Report, Mississippi Agricultural and Forestry Experiment station. 7 (14), 4p.

[107] Sawamoto, T, Kusa, K, Hu, R, & Hatano, R. O, CH_4 and CO_2 emissions from subsurface-drainage in a structured clay soil cultivated with onion in Central Hokkaido, Japan. Soil Sci. Plant Nutr., , 49, 31-38.

[108] Shan, L, & Chen, G. L. (1993). The Theory and Practice of Dryland Agriculture in the Loess Plateau of China, Science Press, Beijing.

[109] Six, J, Ogle, S. M, Breidt, F. J, Conant, R. T, Mosier, A. R, & Paustian, K. (2004). The potential to mitigate global warming with no-tillage management is only released when practiced in the long-term, Global Change Biology, , 10, 155-160.

[110] Six, J, & Paustian, K. (1999). Aggregate and soil organic matter dynamics under conventional and no- tillage systems", Soil Sci. Amer. J. 63(5), 1350-1358.

[111] Smith, P, Martino, D, & Cai, Z. (2008). Greenhouse gas mitigation in agriculture.Philosophical Transactions of the Royal Society, Series B, , 363, 789-813.

[112] Stout, B. A. (1990). Handbook of Energy for World Agriculture. London & New York, Elsevier

[113] Tang, H, Qiu, J, Van Ranst, E, & Li, C. (2006). Estimations of soil organic carbon storage in cropland of China based on DNDC model. Geoderma , 134, 200-206.

[114] Tobert, H. A, Potter, K. N, & Morrison, J. E. Jr: (1998). Tillage intensity and crop residue effects on nitrogen and carbon cycling in a vertisol. Comm. Soil Sci. Plant Analysis., 29, 717-727.

[115] Unger, P. W, Langdale, D. W, & Papendick, R. I. (1988). Role of crop residues- improving, conservation and use. In: Hargrove WL(ED).Cropping strategies for efficient use of nitrogen.Special publication 51. American Society of Agronomy, Medison, Wiscar 100.

[116] Uri, N. D. (2001). The potential impact of conservation practices in US agriculture on global climate change", J. Sust. Ag. 18(1), 109-131.

[117] Van and Kessel, C, Pennock, D. J, & Farrel, R. E. (1993). Seasonal variations in denitrification and nitrous oxide evolution at the land- scape scale. Soil Sci. Soc. Am.J. , 57, 988-995.

[118] Venterea, R. T, Baker, J. M, Dolan, M. S, & Spokas, K. A. (2006). Carbon and nitrogen storage are greater under biennial tillage in a Minnesota corn-soybean rotation. Soil Sci. Soc. Am.J., 70, 1752-1762.

[119] Viek, P. L. G, & Fillery, I. R. P. Improving nitrogen use efficiency in wetland rice soils", Paper red before The Fertilizer Society of London on the 13th December (1984). Fertilizers Society Proceedings (230)

[120] Wang, L G, Qiu, J J, Tang, H J, Hu, L, Li, C S, & Eric, V R. (2008). Modeling soil carbon dynamics in the major agricultural regions of China. Geoderma , 147, 47-55.

[121] Wang, W. J, Moody, P. W, Reeves, S. H, Salter, B, & Dalal, R. C. (2008). Nitrous oxide emissions from sugarcane soils: effects of urea forms and application rate. Proceedings of the Australian Society of Sugar Cane Technologists, , 30, 87-94.

[122] Wanger- Riddle, C, Furon, A, Mclaughlin, N. L, Lee, I, Barbeau, J, Jayasundara, S, Parkin, G, Von Bertoldi, P, & Warland, J. (2007). Intensive measurement of nitrous oxide emissions from a corn-soybean-wheat rotation under two contrasting management systems over 5 years. Global change Biol. , 13, 1722-1736.

[123] Watson, R. T, Zinyowera, M. C, Moss, R. H, et al. (1996). Impacts adaptations and mitigation of climate change: Scientific- technical analyses. Intergovernmental panel on climate change, climate change 1995.Cambridge University press, USA, 879.

[124] Weijin WangDalal RC, Reeves Steven H., Bahl Klaus Butterbach and Kiese Ralf. (2011). Greenhouse gas fluxes from an Australian subtropical cropland under long-term contrasting management regimes, Global Change Biology., 17, 3089-3101.

[125] West, T. O, & Marland, G. (2002). A synthesis of carbon sequestration, carbon emissions and net carbon flux in agriculture: Comparing tillage practices in the United States:, Agr. Ecosyst. Env. , 91, 217-232.

[126] West, T. O, & Post, W. M. (2002). Soil organic carbon sequestration rates by tillage and crop rotation: global data analysis. Soil Sci. Soc. Am. J., 66, 1930-1946.

[127] Wilson, H. M, & Al- Kaisi, M. M. Crop rotation and nitrogen fertilization effect on soil CO_2 emissions in central Iowa, Applied Soil Ecology (2008). , 39(2008), 264-270.

[128] Yao, Z. S, Zheng, X. H, Xie, B, et al. (2009). Tillage and residue management significantly affects N-trace gas emissions during the non-rice season of a subtropical rice-wheat rotation.Soil Biology & Biochemistry, , 41, 2131-2140.

[129] Zadoks, J. C, Chang, T. T, & Konzak, C. F. (1974). A decimal code for the growth stages of cereals. Weed Res. , 14, 415-421.

[130] Zhang, Q Z, Yang, Z L, & Wu, W L. (2008). Role of crop residue management in sustainable agricultural development in the North China Plain. J Sust Agri. , 32, 137-148.

Chapter 8

Major Insects of Wheat: Biology and Mitigation Strategies

Atif Kamran, Muhammad Asif, Syed Bilal Hussain,
Mukhtar Ahmad and Arvind Hirani

Additional information is available at the end of the chapter

1. Introduction

Wheat is one of the major cereal crops with annual global production over 600 MT from about 200 M hectares (FAO 2012). The cultivation of wheat started about 10,000 years ago as part of the Neolithic revolution which state a transition from hunting and gathering of food to settle agriculture. Earlier cultivated forms of wheat were diploid (einkorn) and tetraploid (emmer) with known initial origin of the south-eastern part of Turkey (Dubcovsky and Dvorak, 2007). Subsequent evolutionary adaptation and continuous research produced hexaploid bread wheat that is currently widely adapted in about 95% area of world wheat. Globally, all crop production practices are being highly challeged by biotic and abiotic stresses. Biotic stresses especially insect pests and dieseases causes devastating damage in terms of yield and quality. On average pests cause 20-37% yield losses woldwide which translating to approximately $70 billion annually (Pimentel et al., 1997). In agro-ecosystems, herbivore insects are abudant and likely to colonise within same population and disperse from one crop field to another depending on the availablility of plant tissues and feeding behaviour of insects. Quantitative feeding style of the herbivore insect on specific crop resulting significant damage to the crop during the entire life cyle which is believed specific insect as pest of that perticular crop. Single pest may attack multiple crops within single growing season that make crop rotation and pest management more challenged. Wheat producing areas encounter with either sucking and pericing pests or plant tissue feeding pests. Regional pests also observed in wheat growing areas as major damaging pests woldwide. The breeding strategy againsts these insects/pests heavily rely on the inheritance of resistance mechanism in the crops under consideration. The insect resitance is mainly goverened by three types of mechanisms/genes i.e., oligogenes; where resistance is confered by single genes as in case of hessian fly in wheat, polygenes; where several genes having small and additive effect bring about resistance against insects as in case of cereal leaf beetle in wheat and sometime cytoplasmic genes also confer resitance againsts insects/pest e.g., in maize and lettuce against European corn borer and root aphid respectively.

Large numbers of chemical formulations have been developed as pesticides to chemically control pest problems in different crops, however, control during all stages of insect life i.e. egg, larva, pupa and adult is almost impossible. It is therefore important to understand biology of insect pest simultenously with the crop biology to unserstand when, where and what chemical should be used to control specific insect/pest more effectively. In addition, integrated pest management practices can also enhance control measures with mininum input and with no or less environmental hazards.

In this review, we have outlined major insects of wheat alongwith their biology and controll stretegies to minimize grain yield losses.

2. Wheat aphids

There are six species of aphids that damage cereals.These species include *Rhopalosiphum padi, Schizaphis graminurn, R. Maidis, Metopoliphiurn dirhodum, Sitobion avenae* and *Diuraphis noxia.* Two of the species commonly known as Russian Wheat Aphid (*Diuraphis noxia*)and Bird Cherry-Oat Aphid (*Rhopalosiphum padi*) are considered notorious for their direct and indirect losses.

Russian Wheat Aphid (RWA) is known to be a sporadic insect causing significant yield losses by spreading out from its origin. The centre of origin for RWA is considered to be the central Asian mountains of Caucasus and Tian Shan. The specie could now be found in South Africa, Western United States, Central and Southern Europe and Middle east (Berzonsky et al., 2003). The RWA was first reported in South Africa in 1978 (Walters 1984), in Mexico during 1980 (Gilchrist et al., 1984), in United States in 1986 and Canadian Prairie Provinces during 1988 (Morrison et al., 1988). RWA is present in almost all significant wheat producing areas of the world except Australia (Hughes and Maywald 1990). RWA attacks most of the cereals including wheat, barley, triticale, rye and oat. Alternate hosts for RWA are cool season (crested) and wheat grasses (*Agropyron spp.*). The economic impact of RWA include direct and indirect losses that have been estimated to be $893 million in Western United states during 1987 to 1993 (Morrison and Peairs, 1998) whereas37% yield losses in winter wheat have been reported in Canadian Prairies (Butts et al., 1997).Direct losses have also been assessed as an increased input cost due to insecticides and indirect losses include reduced yield due to RWA infestation.

2.1. Biology

Climatic conditions and temperature in particular, plays a significant role in population dynamics of the aphids. A warmer temperature can potentially accelerate the aphid's growth both in terms of number and size, yet, the extreme temperatures can possibly limit the survival and spread of the aphids. RWA is known to be present in its three different morphological types–immature wingless females, mature wingless females and mature winged females. Winged mature females or *alates* spread the population and infection to the surrounding host plants whereas the wingless types or *apterous* cause damage by curling and sucking the young leaves. Heavily infested plants may typically look prostrated and/or stunted with yellow or whitish streaks on leaves. These streaks, basically, are formed due to the saliva injected by the RWA (Kazemi et al., 2001). The most obvious symptoms in heavy infestations can be reduced leaf area, loss in dry weight index, and poor cholorophyll concentration. Plant losses could be attributed mainly due to reduced photosynthates availability to plants and reduced photo-synthetic activity due to RWA infestation (Millar et al., 1994; Burd and Elliott 1996).The photochemical activity of the plants have been reportedly inhibited by the RWA feeding from leaves and disruption in electron transport chain is likely to be the main cause of the reduced activity (Haile et al., 1999). Spikes can have bleached appearance with their awns tightly held in curled flag leaf. RWA can feed from main stem, flag leaf sheath and/or even developing kernels at flowering, resulting in shrivelled/empty grain or spike death (Peairs 1998a). In the event of sever attack; the wheat tiller can have purplish streaks. Approximately 1% to 0.67% yield losses per percentage of the infested tillers are reported at two tiller stage in Montana and Washington respectively (Archer et al., 1998). Yield losses can greatly vary due to infestation at different growth stages, duration of infestation and climatic conditions (wind patterns and temperature). A number of biotypes for RWA have been reported to be present throughout the cereal production areas of the world. These biotypes are classified due to significant genetic differences among them (Weng et al., 2007).

2.2. Strategies to mitigate RWA

A number of strategies have been deployed to mitigate RWA. Among these strategies, the host plant resistance has been the most effective and economic method to induce antixenosis, antibiosis and/or tolerance against RWA. RWA host plant resistance is well known to be qualitative in nature, and about nine resistance genes have been documented so far. These genes are: *Dn1, Dn2, dn3, Dn4, Dn5, Dn6, Dn7, Dn8, and Dn9* (Du Toit 1989, Nkongolo et al. 1991a, Saidi and Quick 1996, Marais and Du Toit 1993, Marais et al. 1994, Elsidaig and Zwer 1993). A gene-for-gene model supposedly induces resistance against RWA. In this mechanism the resistant gene produces a protein containing nucleotide binding site-leucine rich repeat (NBSLRR) domain (Feuillet et al. 2003, Botha et al. 2005 Jones). This NBSLRR domain first recognizes and then interacts with cognate Avr protein produced by the respective insect (Keen, 1990). Another domain (serine / threonine-protein kinases: STKs) has also been reported to be produced by *Dn* genes to confer resistance against RWA (Boyko et al. 2006). A general practice to introgress resistance in commercial cultivars is a combination of two resistant genes; however there are reports with single resistant gene or a combination of three genes conferring all three types of resistances in small grain cereals. *Dn4* reportedly have been most extensively used gene in breeding for resistant cultivar development (Quick et al., 2001). Rye and common progenitors of wheat (*T. Tauschii* or goatgrass) has served as resistance source for number of genes. *Dn7* gene was introduced in hexaploid wheat through translocation from the rye chromosome 1R to wheat chromosome 1B, and this gene has exhibited the resistance against all the known biotypes of RWA in North America and Africa (Lapitan et al. 2007, and Zaayman et al. 2008). Pyramiding the resistance genes would be ideal to minimize the development of resistant biotype of aphids, yet, at present there is no such differential series of pure lines available to be used as breeding material. Marker assisted selection could be deployed potentially to confer a long term resistance against RWA. A number of alternate methods to control RWA has been suggested and practiced that include cultural, biological and chemical control methods. Cultural control strategies involve eradication of volunteer and alternate host plants is generally recommended. Another strategy is grazing the volunteer plants which significantly reduce the RWA infestation (Walker and Peairs 1998). Adjusting planting dates to de-synchronize the insect population dynamics and favourable environmental conditions of any particular area can also help to control RWA (Butts 1992). The enhanced fertigation of infested field, and biological control of RWA is also possible with 29 different species of insects and 6 fungus species (For further detail the readers are encouraged to read Hopper et al. (1998). Of the predator insects, 4 different species of wasps have become adopted to United States. Besides these cultural practices, chemical control method is also widely practiced with equivocal cost efficiency.

3. Bird cherry-oat aphid

Bird cherry oat aphids can saliently be characterised due to their high adaptive biological plasticity and transmission of viral diseases–*Barley yellow Dwarf* (BYD) virus in particular (Stern 1967). Bird cherry oat aphid is native to almost all over the world (Vickerman and Wratten 1979) and is abundantly found in Northern Europe (Wiktelius, 1982), North America, and New Zealand (Kieckhefer & Gustin, 1967; Kieckhefer, 1975). Bird cherry oat aphid can adopt a number of species as an alternate host including oat, wheat, cereal and other grasses and even on species of families *Juncaceae*or and *Cyperaceae* (Rautapaa, 1970) with primary host being Bird Cherry (*Prunus padus* L.) and closely related tree species. Yield losses caused by Bird cherry oat aphid can vary greatly depending upon the time of infestation in relation to plant growth. It is one of the serious pests of in wheat growing areas of the world due to: a) its longest span of presence from early spring to late autumn (Dedryver 1978), b) ability to overwinter as an egg and/or parthenogentic individuals and c) vectoring the *Barley yellow Dwarf* (BYD) virus.

3.1. Biology

Bird cherry oat aphid has the ability to multiply parthenogenically for one or more than one generation and subsequently undergo sexual reproduction. Bird cherry oat aphid alates fly to the primary host during autumn to mate and produce eggs. Change in environmental conditions stimulates the reproductive growth in Bird cherry oat aphid, to overwinter as eggs (Lees 1966), although it can survive in the regions of mild winter (Carter et al., 1980) and/or by descending down beneath the soil surface and feeding from the base of stalks (Wiktelius, 1987). An equivocal role of temperature in the survival of eggs has been reported in literature with a number of studies reporting the positive correlation between bird cherry oat aphid population and warm winters (Pierre 1987). However, certain clones adaptive to a site of cooler temperatures have shown considerable ability to withstand winter temperatures (Griffiths and Wratten 1979). Therefore, it could be very tempting to conclude a strong positive correlation between temperature and increase in population of Bird Cherry Oat Aphid.

The feeding symptoms of bird cherry oat aphids are almost absent. Direct yield losses caused by bird cherry oat aphid are greatly dependent upon plant growth stage; as 24-65% losses can occur in case of infestation at seedling stage, and very low or non-significant yield losses from booting or later stages have been reported (Kieckhefer et al., 1995; Voss et al., 1997). Indirect yield losses are caused by transmitting viral diseases *e.g.* causing one of the important viral disease, *Barley yellow Dwarf* (BYD), in cereals. Sucking the sap and transmitting the BYD simultaneously can cause even more losses than alone (Riedell, 1999, 2007). The yield losses caused by sucking the plant sap can reduce the grain yield by 15%. The yield losses caused by BYD virus were estimated to be as high as 70% in the individual field of Idaho, with an average loss of 22% in different years depending upon the severity of infestation (Bishop and Sandvol 1984).

3.2. Strategies to mitigate bird cherry oat aphid

Number of studies have produced contrary results in the perspective of host plant resistance against bird cherry oat aphid. This might have happened due to very high biological plasticity of bird cherry oat aphids, presence of number of clones and related species in different geographical regions and different plant traits conferring resistance. Comprehensive and effective resistance against bird cherry oat aphid is typically possible when one has a detailed understanding of plant resistance mechanism to a particular growth stage of bird cherry oat aphid life cycle. In this scenario, numerous experiments have been designed to explore the most effective stage in the life cycle to limit the population of bird cherry oat aphid and its relationship to the extent of plant damage (Rauttapaa 1970; Markkula and Roukka 1972; Lowe 1980). Plant traits or mechanisms that induce nymphal mortality, elongated development at seedling stage and reduce birth rate at flowering are reportedly the most effective mechanisms to manage bird cherry oat aphid (Wiktelius and Pettersson 1985). Plant traits that can prevent the bird cherry oat aphid inoculating the phloem and can reduce the proportional production of winged females, can limit the BYD dispersal to other plants (Gibson and Plumb 1977).

4. Greenbug

Schizaphis graminum Rondani or greenbug is a warm season perennial pest, causing substantial losses to cereal crops and wheat in particular. Greenbug was first reported on oat during early 20th century and also has colonized successfully in sorghum during 1960s (Harvey and Hackerott 1969). Greenbug is known to be originated from Virginia, North America (Hunter 1909), with a contradictory report that it might have originated from Italy (Michuad, 2010). Webster and Amosson (1995) reported 41% dryland and 93% irrigated area under wheat cultivation in Western US was infested with greenbug. A notorious periodic outbreak during 1976 in Oklahoma caused estimated losses exceeding $80 million (Starks and Burton 1977). Large populations of greenbug shift onto sorghum during summer when wheat is harvested and colonize in masses. In absence of sorghum, they can shift to wild grasses which can rarely accommodate larger populations (Anstead et al., 2003).

4.1. Biology

Greenbug is a light green, small size (about 3 mm in length), and sap sucking arthropod. It injects its stylet in sieve tubes, by secreting protenacious saliva to facilitate penetration. Greenbug passively feed on sap upon a successful connection to the sieve tube (Miles 1999). Yellow to red lesions surrounded by a large cholrotic area can be readily identified on leaf surface, which turn necrotic with time. A seven-days feeding of 30 aphids per culm reportedly caused 40% grain weight losses on winter wheat (Kieckhefer and Kantack 1988). Greenbug is also reported to significantly reduce root length (Burton 1986) and hence limiting the plant capability to withstand drought stress. Greenbug has also been confirmed to vector *Barley yellow dwarf virus*.It can multiply asexually year round in a warm area as cold temperatures can significantly limit its survival. Occasional sexual reproduction, supposedly, has lead to the emergence of different biotypes of greenbug, which were eventually identified due to their differential response to resistant cultivars and pesticides (Ullah and Peters 1996; Rider and Wilde 1998). Wood (1961), identified greenbug damage on resistant line DS 28A, and described it as a different biotype which was named as Biotype B. The biotype to which DS 28A was resistant was called Biotype A. Similarly, biotype C was discovered on sorghum due to greenbug substantial damage on cultivar named 'Piper' resistant to biotype B. So far, eleven biotypes has been identified and named biotype A to K. Most prevalent biotypes in Oklahoma and Kansas, which is the area of its most economic threat, are I, E and K (Kindler et al., 2001)whereas biotype G is reported to be more prevalent on non-cultivated grasses in Southern Great Plains (Anstead et al., 2001).

4.2. Strategies to mitigate greenbug

A regular detection of new greenbug biotypes has more or less necessitated the use of two strategies to mitigate its severe outbreaks: the chemical control method and host plant resistance. A number of chemicals have been used against greenbug including dimethoate, parathion, methyl parathion, chlorpyrifos, imidacloprid and malathion with varying doses depending upon the threshold on a specific growth stage. Extensive use of chemicals had not only induced insecticide resistance in the greenbug, but also has environmental concerns in addition to the extra cost. Therefore, the researchers continuously looked for host plant resistance against the greenbug. Qualitative inheritance of resistance conferred by both dominant and recessive genes is well documented in literature with gene symbols as: gb1, Gb2, Gb3, Gb4, Gb5, and Gb6. Gb6 is the most potent gene conferring resistance against biotypes B, C, E, G and I and was recovered from a wheat-rye translocation germplasm by Porter et al., 1994. Theoretically, gene pyramiding could possibly ensure a broad spectrum and long-lasting resistance against greenbug. Porter et al., (2000), designed a study to verify the resistance conferred by more than one gene, and reported no additional protection conferred by more than one gene compared to their single counterpart; and suggested for a sequential release of resistant gene with complete monitoring of greenbug biotypes prevailing in a particular area.

Development of molecular markers flanking these resistant genes is underway to arm the modern molecular techniques to exploit the resistance potential at its maximum.

5. Cereal leaf beetle

Cereal leaf beetle is an insect of cereal or small grain grasses. The particular origin of the insect is still unknown, however, it is considered to be a native insect of Europe and Asia. It is a serious insect in Eastern and South-Eastern Europe including Hungary, Yugoslavia, Poland, and Rumania. It is now considered to be present all over the Europe. In Asia, it is reported to be present in Pakistan, India and Iran. In America, it was probably introduced in early 1960s when it was first identified as a serious insect in Michigan, in 1962. It is now present in most of the states, and in Canadian Prairies–Alberta, Saskatchewan, and Manitoba (Kher et al., 2011). Cereal leaf beetle feeds on oat, wheat, barley in particular and on many other cultivated and non-cultivated grasses (Wilson and Shade 1966). The economic losses caused by cereal leaf beetle greatly vary among the crops, regions and timing and level of infestation. Buntin et al., (2004) reported a maximum loss of 40%; whereas Herbert et al., (2007) reported about 15% wheat yield losses in Virginia due to cereal leaf beetle.

5.1. Biology

Cereal leaf beetle adult is about 5mm long, bluish black head and elytra, and burgundy red thorax and legs. Adult feeding usually does not cause economic losses to the crops. However, the larvae, which is also about 5mm in length and shiny black in colour, feeds on photosynthetic tissues of the leaf, leaving behind the leaf skeleton only (Buntin et al. 2004). This results in significant loss of photosynthetic activity of the plant, giving it a frosted look. Hence, the plant fails to produce expected yield and quality (Merrit and Apple 1966, Grant and Patrick 1993). Cereal leaf beetle generally has one generation per year, however a small second generation is also reported in Virginia (McPherson 1983b). A typical cereal leaf beetle life cycle span is about 46 days, but can be as short as 10 days and as long as 90 days depending upon the environmental conditions and temperature (Guppy and Harcourt 1978, Metcalf and Metcalf 1993). Highest yield losses can be anticipated by the cereal leaf beetle larvae feeding the flag leaf. The losses vary greatly in different regions *e.g.* in Poland the yield losses reported are 3-8% (Ulrich et al., 2004), and could be as high as 95% in The Netherlands (Daamen and Stol 1993) and on different grain crops *e.g.* wheat yield losses in North America can reach up to 55% (Royce 2000), whereas these losses can be 75% in oat and barley (Webster and Smith 1979).

5.2. Strategies to mitigate cereal leaf beetle

Chemical control has long been practiced to control cereal leaf beetle, even before its identification and recognition as a threatening pest. Pesticides have both been applied as granules to soil (Carbofuran) and as a foliar spray (Endosulfon, methomyl, methyl parathion, etc). Non-selective insecticides have indiscreetly killed the natural enemies and the parasitic species. Biological control has also been an effective method to mitigate cereal leaf beetle. A number of species parasitic to larvae and eggs have been reported as *T. julis, Diaparsis carinifer(Thomson) and Lemophagus curtus(Townes) (Hymenoptera: Ichneumonidae), Anaphes flavipes(Foerster) (Hymenoptera: Mymaridae)*(LeSage et al., 2007, Haynes and Gage 1981). Host plant resistance against cereal leaf beetle has been most effective in wheat, mainly due to trichomes (pubescence) produced on leaf surface. A positive correlation between the resistance and trichome length and intensity is reported (Wellso 1973). Non-preferential behaviour for oviposition and first larval instar feeding deterrence are the mechanisms conferring resistance. Oat and Barley have shown lesser resistance against cereal leaf beetle relative to wheat (Hahn 1968). Host plant resistance could not be exploited to its maximum due to variety of reasons: very few resistance sources, lesser adaptation and a negative correlation between resistance and yield are some of them (Kostov 2001).

6. Wheat stem sawfly

The stem sawfly of wheat, *Cephuscinctus* Norton (Hymenoptera: Cephidae), is a phytophagous insect of wheat and other cereal crops including barley, rye and triticale. It is of serious concern in different parts of world especially in northern hemisphere (Shanower and Hoelmer 2001). The *C. cintus* is considered to be a single specie; however differences in virulence have been detected due to genetic variability. Its larvae under different environmental conditions such as similar to North Dakota and Montana differed in duration of post diapause development that might be due to climatic variability. It is one of the major pests of spring wheat in USA. The cropping system like summer fallowing and strip cropping is the main reason to make sawfly as a potential pest causing significant losses. The historical background revealed that *C. cinctus* is indigenous to North America and it exhibits a relationship with Siberian species (Ivie and Zinovjev 1996). Its spread in North America could have occurred due to transport of straw or crown from plants containing live larvae (Ivie 2001). The case of severe infestation of wheat stem sawfly (WSS) was recorded in 1922 in Canada which was due to absence of natural enemies of the sawfly that could result a severe threat to food security. The outbreaks of WSS were short lived because host plants were immediately eliminated due to rust epidemics but the continuous development of rust resistant genotypes lead to progression development of WSS population. Strip farming to control soil erosion is another reason for dissemination of WSS from one field to another.

The biology of WSS revealed that adults of both sexes are weak fliers and cannot fly long distances. The adult feed on exudate moisture and on nectar while resting on plant stem with head in downward position and legs aligned with its body. The life cycle of WSS is synchronised with the phenology of host plant and all growth and development occurs within the host plant except the last stage. The timing of its emergence is greatly influenced by temperature and adults become active during warm season when wind speed is minimum. The cloudy, windy and rainy conditions have an inverse relationship with the activity of WSS. Adult males become visible first as compared to female to ensure mating of females so that most of eggs oviposited in the early flight will be fertilized whereas eggs at the end of flight remained unfertilized. The haploid male will be produced from unfertilized eggs whereas fertilized eggs lead to the development of diploid female. The adults are sexually mature and ready for copulation and oviposition. The female lay 30-50 eggs during her entire life. The egg stage of

WSS consists of 5-7 days in length while larval development last for one month. On completion of developmental phase, the larvae start feeding and filling the stem with excreted plant tissue called frass that ultimately lead to the stem splitting. The larvae then descend down to the base of stem creating a V shape furrow that results in complete cutting of the stem. The larva constructs a thin cellophane structure to get protection. This sealed cocoon help larva to remain protected from environmental hazards and predation.

The protected larvae can survive for months and it passes most of its winter in the crown root since temperature remained higher as compared to ambient temperature. The rate of mortality of larvae becomes high if it is exposed to low temperature. However, pupation occurs if there is rise in temperature and weather is dry. The pupal development depends on climatic conditions like drop in temperature. The pupa is white and as pupal development proceeds, wings start emerging/developing followed by pigmentation in the body that results in a mature adult. The insect remained in soil during winter and under favourable environmental conditions, it emerges out and ready for flight. The distribution of WSS is spatial and temporal. As soon as they emerge from stubbles, they start migrating to the nearby wheat plants. The infestation might be severe if females oviposit first within field margin that often results in the uniform distribution of eggs as the flight progress (Nansen et al. 2005). The release of signalling compounds from plants attracts WSS that often lead to severe infestation. However female is unable to differentiate between damaged and healthier plants.

The mature WSS cause little injury but boring action of larvae is very destructive and is a major cause of severe losses. The declined in phosynthetic activities due to destruction of parenchyma and vascular tissues is one of the main damage caused by larvae. The stem will be hollow in a week as larvae feeds up and down.

The mitigation strategies might include cultural control (strip planting and alternative planting strategies), early forecasting system, simulation modelling for long term planning, biological control, chemical control and development of host plant resistance (gene deployment, resistant cultivar development and cultivar blends). The pheromone monitoring and host-plant semiochemicals techniques could be used as an effective strategy to minimize damage of WSS. However, the future research needs to involve multi scale collaborative efforts among different disciplines to develop a holistic approach to control any outbreak of WSS. Cultural methods are critical to control WSS, therefore, it's important to encourage producers to adopt such procedures which can minimize the WSS population and increase beneficial insects. The use of resistant genotypes having solid stem can contribute to minimize the damage to a greater extent. The use of cultivars blends, IPM (integrated pest management) and ICM (integrated crop management) could be considered as management tools for the control of WSS.

7. Wheat midge

The major pest of spring wheat in most part of world is Wheat midge (WM) which can cause 30% reduction in wheat yield resulting in an economic loss of 30 million dollar. It is also called orange wheat blossom midge and it is the periodic pest of wheat crop in the northern hemisphere and cause significant damage when climatic conditions favours its growth. It's the main pest of China, Europe and North America where winter and spring wheat is being cultivated. WM is serious pest in Canada (Lamb et al. 2000) that has resulted in widespread use of insecticides. The origin of WM was first detected during 1741 in England. The dispersal of WM mainly take place from Europe to North America and then to Asia. Its dissemination is through larvae which remain in the spikes and then stored in the seed after harvesting with combine harvester. WM hibernate in the soil and during spring season it multiply and pupate. The hatching of cocoon depends on soil temperature and moisture that result in higher numbers. At the ear emergence, the adult WM mates and females then move to wheat crop where it starts laying eggs. The flight of females takes place at evening and if wheat crop is absent laying of eggs take place at barley or weed grasses. The hatching of larvae from eggs takes place after a week and produces alpha-amylase enzyme to release sugars from the grain. The larvae then drop to the soil after feeding for few weeks and made a cocoon around itself. The generation of WM completes in one year and it passes winters in soil as larvae. The high temperature terminates the diapause of larvae and it comes out from cocoons and spends some time at soil surface (Doane and Olfert, 2008). The damage to the crop starts at grain development stages causing shrivelling and crack which ultimately reduces yield and quality of crop.

The development of WM is highly dependent on soil moisture and temperature. The termination of larval diapause occurred in phases: firstly, larvae required cool temperature for three months; secondly, larvae enter into moisture sensitive phase which remained for 5-6 weeks. However, if soil is dry it remained in diapause for one year while on the other hand if moisture is sufficient, larvae's terminated diapause, pupated and emerged as adults within five weeks. The adult's stage is last stage of WM and basically it is small orange fly with length of 2-3mm. It has two large black eyes with size equivalent to mosquito and has three pairs of legs which are larger in size. The wings are oval shaped and transparent. The adults will prefer to remain in crop canopy where the environment is humid and when conditions become favourable the female become active and comes at the top of canopy starting laying eggs on newly emerged spike. Therefore, WM larvae compete directly with humans for the grain and destroy the grain by causing shrivelling. The infestation of WM can be seen on all parts of spike and feeding of larvae is greater on small seeds as compared to larger one (Lamb et al. 2000). The intensity of damage could be determined from the feeding place of larvae. If it feeds closer to the grain embryo, the attack will be very severe. Usually, the seed is attacked by larger number of larvae but if four or more is present attack will be of serious nature. The body size of larvae might be affected significantly if they are present on one seed because of competitions between them. The damage caused by larvae to the wheat seed can be calculated by dividing mass lost by the seed to the mass gain by the larvae (Lamb et al. 2000) called as efficiency index of WM larvae. The activity of WM larvae decreases when wheat seeds have lost 75% of their mass. WM feeding has resulted in maximum impact among feeding insects that feed on crops belong to Poaceae family (Gavloski and Lamb 2000). The damage of WM to crop adversely affects the agronomic performance like resistance to sprouting, yield, germination and seedlings early vigor. It also affects grain quality resulting to change in seed protein levels and decline in the drought resistance patterns of crop. The quality might be further deteriorated due to carrying of harmful microorganism with WM and attack by the semolina after the WM.

The WM could be controlled by inspecting field at heading stage and by the application of insecticide to minimize the damage. If infestation of WM is identified at early stage by regular monitoring at heading and flowering then WM attack could be minimized to a greater extent. The use of wheat genotypes that are resistant to WM is another way to control its attack. However, it has been recommended that the best control measure is to use predators that can eat the WM larvae so that it is unable to multiply further. The examples of predators include polyphagpus which might control WM at the different vulnerable stages. The concept of host plant resistance is another way to control WM by developing such genotypes that can resist the development of WM. The host plant resistance includes resistance mechanism and genetics in which genotypes produce antitoxic substances lead to minimize WM infestation. These genotypes changes oviposition in the field and reduce the egg densities in the field resulting in lower numbers of WM. The research studies has depicted that these lines could control WM larvae from 58 to 100% (Lamb et al. 2000).The development of antibiosis is another way to control WM and resistance in spring wheat is linked with the production of phenolic compounds from seeds which might destroy the WM (Ding et al. 2000).In the same way, use of selection protocols and field methods like screening of wheat genotypes and cultural practices are the important ways to control WM. The modifications in the oviposition sites can also control WM to a considerable degree. Breeding wheat for resistance to insects is an easiest and cheapest mean to control WM

8. Hessian fly

The *Mayetiola destructor* called Hessian fly (HF) belongs to the species of fly and is destructive pest of cereal crops including wheat, barley and rye. It is native of Asia and transported to Europe and North America through straw. HF has two generations in a year but it can go to five. The dark coloured female lays 250-300 reddish eggs on plants during spring season. After 3-10 days larvae hatch from the eggs and they cannot survive in the open air therefore they move to the base of leaf sheath which is preferred feeding site. The larvae (maggots) crawl down to the crown of the plant during fall season. The meristemic activities in node causes wheat stem to elongate and maggots are usually found at the top of leaf nodes. The HF infestation will be found at the top because female prefer to lay eggs on new leaves which comes out from nodes. The maggots are reddish brown and as they feed and grow it changes colour become white and greenish white. The feeding of maggots is on stem and after scraping the stem it start sucking up the sap which comes out from the wound. The larvae remain feeding for fourteen to thirty days. The flaxseed is the shiny, protective case where maggot spent its last stage and it is built from insect skin and has resemblance with the seed of flax plant. The attack of larvae is so severe that plants are unable to bear grain. The HF comes out from the flaxseed structure when climatic conditions become favourable. The adults come out and start new generations and if climatic conditions are extreme (too hot or cold) it remained inside the flaxseed coat until climatic conditions become favourable. The presence of HF and there maggots at the same time is very uncommon particularly during heavy infestations. The complete life cycle from egg to adults requires 35 days if temperature is favourable.

The damage caused by HF maggots is mainly on vegetative growth which might reduce the activity of photosynthesizing machinery resulting to stunting growth. The maggots during feeding also inject toxic substances resulting to inhibition of plant growth. These toxin acts as inhibitors to the plants and overall hormonal action of plants disturbs resulting to poor growth. However damage could be severe if timing and degree of infestation is perfectly matched with crop phenological stages. The single maggots can cause significant damage to wheat plant because toxins released during feeding interfere with wheat crop growth. Meanwhile if the attack of larvae is at single leaf stage then it will be killed immediately. The attack at later stages cause destruction of first tillers and growth of the crop delayed. The weakening and shortening of stem occur due to larvae attack and it might break from the first or second node before the harvest of crop resulting to head loss. The reduction in yield and quality of crop will be observed with severe mechanical losses to stem and head during heavy infestation.

The use of preventive rather than chemical control methods can control the population dynamics of the insect. These methods include biological and cultural approaches which are reliable and feasible for wheat crop growers. The growing of resistance cultivars by the use of biotechnology is the best way to control the damaged caused by HF. The tissues of plants contain several types of carbohydrate binding proteins called lectins. These lectins have potential to build resistance in the wheat against insects. The identification of genes which might produce this type of lectins will be best way to control HF. The genes includes Hfr-2 called as HF destructor which is expressed in the leaf sheaths of the resistance genotypes (Puthoff et al., 2005). Similarly mannose binding lectins which serve as storage protein accumulates in the phloem sap and might act against HF. These lectins have anti insect properties because it accumulates in the midgut of insect and kill them immediately. The production of Wci-1 mRNAs and Hfr-1 in response to the attack of HF larvae is another defensive mechanism which is present in resistant varieties of wheat. The Hfr-1 gene is called defender gene against HF and it can control crop from severe attack (Subramanyam et al., 2006). Meanwhile there are number of different sources of incorporation of resistance traits into wheat which might built defensive mechanism in crop against HF. Antibiosis is the main mechanism of resistance associated with these genes and is expressed as the death of first larvae. The biochemical nature of antibiosis in wheat includes development of silica in sheaths and production of free amino acids, organic acids and sugars in plants. The development of resistance genotypes in wheat breeding programme might improve the durability of resistance in wheat genotypes against HF. The breeding programmes include use of resistant genes or combination of different level of resistance genes that might respond differential to abiotic and biotic stresses. The use of genes which have potential to control HF is best way to control population dynamics of HF larvae in wheat crop. The knowledge of molecular markers and QTL mapping associated with resistance genes incorporation in wheat is another landmark which might be used to control HF. The use of simulation genetic models might be used to check the development of single gene resistance in crop and it is adequate way to control the HF.

The HF population dynamics could be controlled by modification in tillage practices and change in the cropping pattern which can destroy the life cycle of pest. The delayed planting is another way to control the HF. There are large numbers of different parasitoids which attack the HF and might be used to control its attack on crop. The use of chemical to control HF is not recommended. The best way to control HF is development of resistant genotypes which work like systemic insecticides. Similarly production of novel jacalin like lectin gene from wheat responds significantly to the infestation of HF larvae and could be use effectively in future breeding programmes. The wheat genotypes having higher levels of Hfr-1 at the larval feeding sites and only small amount of lectin at these sites will control the larvae.

Insect	Resistant Gene	Primer Sequences	Gene Origin	Gene Location	Affiliation
GreenBug	gb1	Not mapped	T. Turgidum/T. Durum		
	Gb2	ATATCTCAACCAACTTCACAAAGTC CATTGTTTAAAAAGAGGGGATATG	S. Cereale		Lu et al., 2010
	Gb3	5'- AGC GAG GAG GAT GCA TCT TAT T -3' 5'- GAC ATA CAC ATG ATG GAC ACG G-3'	T. Tauschii	7DL	Weng et al., 2002.
	Gb4	Not mapped	T. Tauschii		
	Gb5	Not mapped	T. Speltoides		
	Gb2/Gb6	TATACACCAACAAGTAGCGACAATA AAACAAACCTTCAGTATCTTCTCAC	S. Cereale		Lu et al., 2010
	RAPD-PCR based Single Decamers	5'-CTCACCGTCC-3'			Kharrat et al., 2012
		5'- GAGCCCTCCA-3'			Kharrat et al., 2012
		5'-TCACGTCCAC-3'			Kharrat et al., 2012
		5'- GGCTCATGTG-3'			Kharrat et al., 2012
		5'- AGTC-GTCCCC-3'			Kharrat et al., 2012
Hesian Fly	H9	5'- GGA AGC GCG TCA GCA CTA GGC AAC -3' 5'- GGC TTC TAG GTG CTG CGG CTT TTG TC -3'	T. Aestivum	1AS	Kong et al., 2005
	H13	5'- CAA ATG CTA ATC CCC GCC -3' 5'- TGT AAA CAA GGT CGC AGG TG -3'	T. Aestivum	6D	Liu et al., 2005
	H25	5'- CTG CCT TCT CCA TGG TTT GT -3' 5'- AAT GGC CAA AGG TTA TGA AGG -3'	S. Cereale	4A	Sebesta et al., 1997
	H26/H32	5'- CCT AAC TGA GGT CCC ACC AA -3' 5'- GCA AAG GAC TTG ATG CCT GT -3'	T. Tauschii		Yu et al., 2010
	H31	5'- TCC TAC CTC CAT TCC CCT TT -3' 5'- TCA AAA TGA ATC GGA AGG GT -3'	T. Turgidum	5BS	Williams et al., 2010
	Hdic	5'- GAC AGC ACC TTG CCC TTT G -3' 5'- CAT CGG CAA CAT GCT CAT C -3'	T. turgidum ssp. dicoccum	1AS	Liu et al., 2005
Stem Saw Fly	Qss.msub-3BL	5'- GCA ATC TTT TTT CTG ACC ACG -3' 5'- ACG AGG CAA GAA CAC ACA TG -3'		3BL	Cook et al.,
	Qss.msub-3BL	5' GCAATCTTTTTTCTGACCACG 3' 5' ATGTGCATGTCGGACGC 3'	Durum wheat	3BL	

Insect	Resistant Gene	Primer Sequences	Gene Origin	Gene Location	Affiliation
RWA	Qss.msub-3BL	5' GTTGTCCCTATGAGAAGGAACG 3' 5' TTCTGCTGCTGTTTTCATTTAC 3'		3BL	
	Dn1	5' TCTGTAGGCTCTCTCCGACTG 3' 5' ACCTGATCAGATCCCACTCG 3'	T. Aestivum		Peng et al., 2007
	Dn2	5'- GAT CAA GAC TTT TGT ATC TCT C -3' 5'- GAT GTC CAA CAG TTA GCT TA -3'	T. Aestivum	7D/1B	Peng et al., 2007
	dn3	not mapped		.	.
	Dn4	5'-CTG TTC TTG CGT GGC ATT AA-3' 5'-AAT AAG GAC ACA ATT GGG ATG G-3'	T. Aestivum	1DS	Peng et al., 2007
	Dn5	5' TCTGTAGGCTCTCTCCGACTG 3' 5' ACCTGATCAGATCCCACTCG 3'	T. Aestivum	7DS	Peng et al., 2007
	Dn6	5' TCTGTAGGCTCTCTCCGACTG 3' 5' ACCTGATCAGATCCCACTCG 3'	T. Aestivum	7DS	Peng et al., 2007
	Dn7	Xscb241 RFLP marker		1RS	
	Dn8	5' TTCCTCACTGTAAGGGCGTT 3' 5' CAGCCTTAGCCTTGGCG 3'	T. Aestivum	7DS	Peng et al., 2007

Table 1. Resistant genes for different insects along with their primer sequences, origin, and location

Author details

Atif Kamran[1], Muhammad Asif[2*], Syed Bilal Hussain[3], Mukhtar Ahmad[4] and Arvind Hirani[5]

*Address all correspondence to: akamran1@ualberta.ca

1 Seed Centre, Department of Botany, University of the Punjab, Lahore, Pakistan

2 Agricultural, Food and Nutritional Science, 4-10 Agriculture/Forestry Centre, University of Alberta, Edmonton, AB, Canada

3 Faculty of Agricultural Sciences & Technology, Bahauddin Zakariya University, Multan, Pakistan

4 Department of Agronomy, PMAS Arid Agriculture University Rawalpindi, Pakistan

5 Department of Plant Science, University of Manitoba, Winnipeg, Manitoba, Canada

References

[1] Anstead, J. A, Burd, J. D, & Shufran, K. A. (2003). Over-summering and biotypic diversity of *Schizaphis graminum* (Homoptera: Aphididae) populations on noncultivated grass hosts. Environ Entomol , 32, 662-667.

[2] Anstead, J. A. (2000). Genetic and biotypic diversity of greenbug (*Schizaphis graminum* Rondani) populations on non-cultivated hosts. M.S. Thesis. Okla. State Univ., Stillwater, OK.

[3] Archer, T. L, Peairs, F. B, Pike, K. S, Johnson, G. D, & Kroening, M. (1998). Economic injury levels for the Russian wheat aphid (Homoptera: Aphididae) on winter wheat in several climate zones. J. Econ. Entomol. , 91, 741-747.

[4] Berzonsky, W. A, Herbert, W. O, Patterson, F. L, Ding, H, Peairs, F. B, Haley, S. D, Porter, D. R, Harris, M. O, Ratcliffe, R. H, & Lamb, R. J. RIH Mckenzie, and TG Shanower. (2003). Breeding wheat for resistance to insects. Plant Breeding Reviews. , 22

[5] Bishop, G. W, & Sandvol, L. (1984). Effects of barley yellow dwarf on yield of winter wheat. Abstracts of Reports of the 43rd Annual Pacific North West Vegetable Insect Conference., 28.

[6] Botha, A-M, Lacock, L, Van Niekerk, C, Matsioloko, M. T, De Preez, F. B, Loots, S, Venter, E, Kunert, K. J, & Cullis, C. A. (2005). Is photosynthetic transcriptional regulation in *Triticum aestivum* L. cv. ÔTugelaDN_ a contributing factor for tolerance to *Diuraphis noxia* (Homoptera: Aphididae)? Plant Cell Rep. , 25, 41-54.

[7] Boyko, E. V, Smith, C. M, Thara, V. K, Bruno, J. M, Deng, Y, Strakey, S. R, & Klaahsen, D. L. (2006). Molecular basis of plant gene expression during aphid invasion: wheat *Pto-* and *Pti*-like sequences are involved in interactions between wheat and Russian wheat aphid (Homoptera: Aphididae). J. Econ. Entomol. , 99, 1340-1445.

[8] Buntin, G. D, Flanders, K. L, Slaughter, R. W, & Delamar, Z. D. (2004). Damage loss assessment and control of the cereal leaf beetle (Coleoptera: Chrysomelidae) in winter wheat. Journal of Economic Entomology. 97, 374-382.

[9] Burd, J, & Elliott, N. (1996). Changes in chlorophyll a fluorescence induction kinetics in cereals infested with Russian wheat aphid (Homoptera: Aphididae). J. Econ. Entomol. , 89, 1332-1337.

[10] Burton, R. L. (1986). Effect of greenbug (Homoptera: Aphididae) damage on root and shoot biomass of wheat seedlings. J. Econ. Entomol. , 79, 633-636.

[11] Butts, R. A. (1992). The influence of seeding dates on the impact of fall infestations of Russian wheat aphid in winter wheat. In: W. P. Morrison (ed.), Proc. Fifth Russian-Wheat Aphid Conference. Great Plains Agr. Council Publ. 142., 120-123.

[12] Butts, R. A, Thomas, J. B, Lukow, O, & Hill, B. D. (1997). Effect of fall infestations of Russian wheat aphid (Homoptera: Aphididea) on winter wheat yield and quality on the Canadian Prairies. J. Econ. Entomol. , 94, 1005-1009.

[13] Carter, N, Mclean, I. F. G, Watt, A. D, & Dixon, A. F. G. (1980). Cereal aphids-a case study and review.-*Appl. Biol. 5*, 271-348.

[14] Cook, J. P, Wichman, D. M, Martin, J. M, Bruckner, P. L, & Talbert, L. Identification of microsatellite markers associated with a stem solidness locus in wheat. E. Crop Science , 44, 1397-1402.

[15] Daamen, R. A, & Stol, W. (1993). Surveys of cereal diseases and pests in the Netherlands. 6. Occurrence of insect pests in winter wheat. Neth. J. Pl. Path. , 99, 51-56.

[16] Dedryver, C. A. (1978). Biologie des pucerons des cer£ales dans l'ouest de la France. 1.-Repartition et evolution des populations de *Sitobion avenae* F., *Metopolophium dirhodum* Wlk., et *Rhopalosiphum padi* L., de 1974 a 1977 sur ble d'hiver dans le bassin de Rennes.-*Ann.ZooL, Ecol. Anirn. 10*, 483-505.

[17] Ding, H, Lamb, R. J, & Ames, N. (2000). Inducible production of phenolic acids in wheat and antibiotic resistance to *Sitodiplosismosellana*. J. Chem. Ecol. , 26, 969-985.

[18] Doane, J. F, & Olfert, O. (2008). Seasonal development of wheat midge, Sitodiplosis mosellana (Ge´ hin) (Diptera: Cecidomyiidae), in Saskatchewan, Canada. Crop Prot. , 27, 951-958.

[19] Du ToitF. (1989). Inheritance of resistance in two *Triticum aestivium* lines to Russian wheat aphid (Homoptera: Aphididea). J. Econ. Entomol. , 82, 1251-1253.

[20] Dubcovsky, J, & Dvorak, J. (2007). Genome plasticity: a key factor in the success of polyploid wheat under domestication. Science 2007;, 316, 1862-1866.

[21] Elsidaig, A. A, & Zwer, P. K. (1993). Genes for resistance to Russian wheat aphid in PI 294994 wheat. Crop Sci. , 33, 998-1001.

[22] FAO ((2012). Food and agricultural organization: Global wheat cultivation areas and production (http://wwwfao.org/index_en.htm)

[23] Feuillet, C, Travella, S, Stein, N, Albar, L, Nublat, A, & Keller, B. (2003). Map-based isolation of the leaf rust disease resistance gene *Lr10* from the hexaploid wheat (*Triticum aestivum* L.) genome. Proc. Natl. Acad. Sci. U.S.A. , 100, 15253-15258.

[24] Gavloski, J. E, & Lamb, R. J. (2000). Specific impacts of herbivores: Comparing diverse insect species on young plants. Environ. Entomol. , 28, 1-7.

[25] Gibson, R. W, & Plumb, R. T. (1977). Breeding plants for resistance to aphid infestation. *In* K.F. Harris and K. Maramorosch (ed.) Aphids and virus vectors. Academic Press, New York, NY., 473-500.

[26] Gilchrist, L. I, Rodriguez-montessoro, R, & Burnett, P. A. (1984). The extent of Freestate streak and *Diuraphis noxia* in Mexico. In: P. A. Burnett (ed.), Barley Yellow

Dwarf, A Proceedings of a Workshop, CIMMYT, Mexico, December 6-8, 1983. CIMMYT, Mexico., 157-163.

[27] Grant, J. F, & Patrick, C. R. (1993). Distribution and seasonal phenology ofcereal leaf beetle (Coleoptera: Chrysomelidae) on wheat in Tennessee.Journal of Entomological Science. *28*, 363-369.

[28] Griffiths, E, & Wratten, S. D. (1979). Intra- and inter-specific differences in cereal aphid low-temperature tolerance. *Entomologia Experimentalis et Apphcata 26*, 161-167.

[29] Guppy, J. C, & Harcourt, D. G. (1978). Effects of temperature on development of the immature stages of the cereal leaf beetle, *Oulema melanopus*. Canadian Entomologist. *10*, 257-63.

[30] Rudd, H. L. U, J. C, Burd, J. D, & Weng, Y. (2010). Molecular mapping of greenbug resistance genes Gb2 and Gb6 in T1AL.1RS wheat-rye translocations. Plant Breeding , 129, 472-476.

[31] Hahn, S. K. (1968). Resistance of barley (*Hordeum vulgare* L. Emend. Lam.) to cereal leaf beetle (*Oulema melanopus* L.). Crop Sci. , 8, 461-464.

[32] Haile, F. J, Higley, L. G, Ni, X, & Quisenberry, S. S. (1999). Physiological and growth tolerance in wheat to Russian wheat aphid (Homoptera: Aphididae) injury. Environ. Entomol. , 28, 787-794.

[33] Harvey, T. L, & Hackerott, H. L. (1969). Recognition of a greenbug biotype injurious to sorghum. J. Econ. Entomol. , 62, 776-779.

[34] Haynes, D. L, & Gage, S. H. (1981). The cereal leaf beetle in North America. Ann. Rev. Entomol. , 26, 259-287.

[35] Helgesen, R. G. (1969). The within generation population dynamics of the cereal leaf beetle, Oulema melanopus (L.), Ph.D. dissertation. Michigan State University, East Lansing, Michigan.

[36] Helgesen, R. G, & Haynes, D. L. (1972). Population dynamics of the cereal leaf beetle Oulema melanopus (Coleoptera Chrysomelidae) a model for age specific mortality. Canadian Entomologist. , 104, 797-814.

[37] Herbert, D. A, Jr, J. W, Van Duyn, M. D, & Bryan, J. B. Karren. (2007). Cereal Leaf Beetle, *In* G. D. Buntin, K. S. Pike, M. J. Weiss, and J. A. Webster (eds.), Handbook of small grain insects. Entomological Society of America, Lanham, MD, 120.

[38] Hopper, K. R, Coutinot, D, Chen, K, Kazmer, D. J, Mercadier, G, Halbert, S. E, Miller, R. H, Pike, K. S, & Tanigoshi, L. K. (1998). Exploration for natural enemies to control *Diuraphis noxia* (Homoptera: Aphididae) in the United States. In: S. S. Quisenberry and F. B. Peairs (eds.), A response model for an introduced pest-the Russian wheat aphid. Thomas Say Publ. Entomol., Entomol. Soc. Am., Lanham, MD., 166-182.

[39] Hughes, R, & Maywald, G. (1990). Forecasting the favourableness of the Australian environment for the Russian wheat aphid, *Diuraphis noxia* (Homoptera: Aphididae), and its potential impact on Australian wheat yields. Bul. Entomol. Res. , 80, 165-175.

[40] Hunter, S. J. (1909). The greenbug and its enemies. Univ. Kans. Bul. , 9, 1-163.

[41] Ivie, M. A. (2001). On the geographic origin of the wheat stem sawfly (Hymenoptera: Cephidae): a new hypothesis of introduction from northeastern Asia. American Entomologist, , 47, 84-97.

[42] Ivie, M. A, & Zinovjev, A. G. (1996). Discovery of the wheat stem sawfly (Cephus-cinctus Norton) (Hymenoptera: Cephidae) in Asia, with the proposal of a new synonymy. The Canadian Entomologist, doi:10.4039/Ent128347-2., 128, 347-348.

[43] Kazemi, M. H, Talebi-chaichi, P, Shakiba, M. R, & Jafarloo, M. M. (2001). Biological responses of Russian wheat aphid, Diuraphis noxia (Mordvilko) (Homoptera: Aphididae) to different wheat varieties. J. Agric. Sci. Technol. , 3, 249-255.

[44] Keen, N. T. (1990). Gene-for-gene complementarity in plantpathogen interactions. Annu. Rev. Genet. 24: 425Ð429.

[45] Kharrat, I, Bouktila, D, Khemakhem, M. M, Makni, H, & Makn, M. (2012). Biotype characterization and genetic diversity of the greenbug,

[46] Kher, S. V, Dosdall, L. M, & Carcamo, H. A. (2011). The cereal leaf beetle: Biology, distribution, and prospects for control. Prairie soil and crop science Journal , 4

[47] Kieckhefer, R. W, & Gustin, R. D. (1967). Cereal aphids in South Dakota. I. Observations of autumnal bionomics.-*Ann. ent. Soc. Am. 60*, 514-516.

[48] Kieckhefer, R. W, & Kantack, B. H. (1988). Yield losses in winter grains caused by cereal aphids (Homoptera: Aphididae) in South Dakota. J. Econ. Entomol. , 81, 317-321.

[49] Kieckhefer, R. W. (1975). Field populations of cereal aphids in South Dakota spring grains.-*J. econ. Ent. 68*, 161-164.

[50] Kindler, S. D, Harvey, T. L, Wilde, G. E, Shufran, R. A, Brooks, H. L, & Sloderbeck, P. E. (2001). Occurrence of greenbug biotype K in the field. J. Agric. Urban. Entomol. , 18, 23-34.

[51] Kong, L, Ohm, H. W, Cambron, S. E, & Williams, C. E. (2005). Molecular mapping determines that Hessian fly resistance gene H9 is located on chromosome 1A of wheat. Plant Breeding , 124, 525-531.

[52] Kostov, K. (2001). Breeding wheat lines for host-plant resistance to cereal leaf beetle by using the cross mutation method. Bulgarian J. Agric. Sci. , 7, 7-14.

[53] Lamb, R. J, Tucker, J. R, Wise, I. L, & Smith, M. A. H. (2000). Trophic interaction betweenSitodiplosismosellana(Diptera: Cecidomyiidae) and spring wheat: implications for seed production. Can. Entomol. , 132, 607-625.

[54] Lapitan, N. L. V, Peng, J, & Sharma, V. (2007b). A high density map and PCR markers for Russian wheat aphid resistance gene*Dn7*on chromosome 1RS/1BL. Crop Sci. , 47, 811-820.

[55] Lees, A. D. (1966). The control of polymorphism in aphids. *Advances in Insect Physiology 3*, 207-277.

[56] LeSageL., E.J. Dobesberger, and C.G. Majka. (2007). Introduced leaf beetles of Maritime Provinces, 2: The cereal leaf beetle *Oulema melanopus* (Linnaeus) (Coleoptera: Chrysomelidae). Proc. Entomol. Soc. Wash. , 109, 286-294.

[57] Liu, X. M, Brown-guedira, G. L, & Hatchett, J. H. Owuoche, Chen MS. (2005). Genetic characterization and molecular mapping of a Hessian fly-resistance gene transferred from *T. turgidum* ssp.*dicoccum* to common wheat. Theoretical and Applied Genetics, , 111, 1308-1315.

[58] Liu, X. M, Gill, B. S, & Chen, M. S. (2005). Hessian fly resistance gene H13 is mapped to a distal cluster of resistance genes in chromosome 6DS of wheat.. In: Theoretical and Applied Genetics , 111, 243-249.

[59] Marais, G. F. and F. Du Toit. (1993). A monosomic analysis of Russian wheat aphid resistance in the common wheat PI 294994. Plant Breed. , 111, 246-248.

[60] Marais, G. F, & Horn, M. and F. Du Toit. (1994). Intergeneric transfer (rye to wheat) of a gene(s) for Russian wheat aphid resistance. Plant Breed. , 113, 265-271.

[61] Markkula, M, & Rautapaa, J. (1963). PVC rearing cages for aphid investigations.- *Annls agric. fenniae 2*, 208-211.

[62] Mcpherson, R. M. (1983b). Seasonal abundance of cereal leaf beetles (Coleoptera: Chrysomelidae) in Virginia small grains and corn. Journal of Economic Entomology. *76*, 1269-1272.

[63] Merritt, D. L, & Apple, J. W. (1969). Yield reduction of oats caused by thecereal leaf beetle. Journal of Economic Entomology. , 62, 298-301.

[64] Metcalf, R. L, & Metcalf, R. A. (1993). Destructive and useful insects: their habits and control, 5th ed. McGraw-Hill, Inc., New York

[65] Michuad, J. P. (2010). Implication of climate change for cereal aphids on the great plains of North America. P. Kindlmann et al. (eds.), Aphid Biodiversity under Environmental Change, 69 DOI

[66] Miles, P. W. (1999). Aphid saliva. Biol. Rev. , 74, 41-85.

[67] Miller, H, Porter, D, Burd, J, Mornhinweg, D, & Burton, R. (1994). Physiological effects of Russian wheat aphid (Homoptera: Aphididae) on resistant and susceptible barley. J.Econ. Entomol. , 87, 493-499.

[68] Morrison, W. P, & Peairs, F. B. (1998). Introduction: response model concept and economic impact. In: S. S. Quisenberry and F. B. Peairs (eds.), A response model for an

introduced pest-the Russian wheat aphid. Thomas Say Publ. Entomol., Entomol. Soc. Am., Lanham, MD., 1-11.

[69] Morrison, W, Baxendale, F, Brooks, L, Burkhardt, C, Campbell, J, Johnson, G, Massey, W, Mcbride, D, Peairs, F, & Schultz, J. (1988). The Russian wheat aphid: a serious new pest of small grains in the Great Plains. Great Plains Agricultural Council Pub. 124, 5 p.

[70] Nansen, C, Macedo, T. B, Weaver, D. K, & Peterson, R. K. D. (2005). Spatiotemporal distributions of wheat stem sawfly eggs and larvae in dryland wheat fields. The Canadian Entomologist, doi:10.4039/N04-094., 137, 428-440.

[71] Nkongolo, K. K, Quick, J. S, Limin, A. E, & Fowler, D. B. and inheritance of resistance to Russian wheat aphid in *Triticum* species, amphiploids and *Triticum tauschii*. Can. J. Plant Sci. , 71, 703-708.

[72] Peairs, F. B. (1998a). Aphids in small grains. Colorado State Univ. Service in Action FactSheet 5.568.

[73] Peairs, F. B. (1998b). Cultural control tactics for management of the Russian wheat aphid (Homoptera: Aphididae). In: S. S. Quisenberry and F. B. Peairs (eds.), A response model for an introduced pest-the Russian wheat aphid. Thomas Say Publ.Entomol., Entomol. Soc. Am., 288-296.

[74] Pierre, J. S. (1987). Investigation of explanatory climatic variables with a view to forecasting outbreaks of insects. Utilization of a delayed integral correlation method. Bull SROP , 10, 109-118.

[75] Pimentel, D, Houser, J, Preiss, E, White, O, Fang, H, Mesnick, L, Barsky, T, Tariche, S, Schreck, J, & Alpert, S. (1997). Water resources: agriculture, the environment, and society. *BioScience.47*(2), 97-106.

[76] Porter, D. R, & Webster, J. A. (2000). Russian wheat aphid-induced protein alterations in spring wheat. Euphytica , 111, 199-203.

[77] Porter, D. R, Friebe, B, & Webster, J. A. (1994). Inheritance of greenbug biotype G resistance in wheat. Crop Sci. , 34, 625-628.

[78] Prokrym, D. R, Pike, K. S, & Nelson, D. J. (1998). Biological control of *Diuraphis noxia* (Homoptera: Aphididae): implementation and evaluation of natural enemies. In: S. S. Quisenberry and F. B. Peairs (eds.), A response model for an introduced pest-the Russian wheat aphid. Thomas Say Publ. Entomol., Entomol. Soc. Am., Lanham, MD., 183-208.

[79] Puthoff, D. P, Sardesai, N, Subramanyam, S, Nemacheck, J. A, & Williams, C. E. (2005). Hfr-2, a wheat cytolytic toxin-like gene, is upregulated by virulent Hessian fly larval feeding. Mol. Plant Pathol. , 6, 411-423.

[80] Quick, J. S, Stromberger, J. A, Clayshulte, S, Clifford, B, Johnson, J. J, Peairs, F. B, Rudolph, J. B, & Lorenz, K. (2001). Registration of 'Prowers' wheat. Crop Sci. , 41, 928-929.

[81] Rautapaa, J. (1970). Preference of cereal aphids for various cereal varieties and species of Gramineae, Juncaceae and Cyperaceae.-*Annls agric. fenniae 9*, 261-211.

[82] Rider SD JrWilde GE ((1998). Variation in fecundity and sexual morph production among insecticide-resistant clones of the aphid *Schizaphis graminum* (Homoptera: Aphididae). J Econ Entomol , 91, 388-391.

[83] Riedell, W. E, & Blackmer, T. M. (1999). Leaf reflectance spectra of cereal aphid-damaged wheat. Crop Sci. , 39, 1835-1840.

[84] Riedell, W. E, Kieckhefer, R. W, Haley, S. D, Langham, M. A. C, & Evenson, P. D. (1999). Winter wheat responses to bird cherry-oat aphid and barley yellow dwarf virus infection. Crop Sci. , 39, 158-163.

[85] Riedell, W. E, Osborne, S. L, & Jaradat, A. A. (2007). Crop mineral nutrient and yield responses to aphids or barley yellow dwarf virus in spring wheat and oat. Crop Sci. , 47, 1553-1560.

[86] Saidi, A, & Quick, J. S. (1996). Inheritance and allelic relationships among Russian wheat aphid resistance genes in winter wheat. Crop Sci. , 36, 256-258.

[87] Schizaphis graminum (Hemiptera: Aphididae)in north Tunisia. Revista Colombiana de Entomología , 38(1), 87-90.

[88] Sebesta, E. E, Hatchett, J. H, Friebe, B, Gill, B. S, Cox, T. S, & Sears, R. G. (1997). Registration of KS92WGRC17, KS92WGRC18, KS92WGRC19, and KS92WGRC20 winter wheat germplasms resistant to Hessian fly.. In: Crop Science, 1997, 37(2):635.

[89] Shanower, T. G, & Hoelmer, K. A. (2004). Biological control of wheat stem sawflies: past and future. Journal of Agricultural and Urban Entomology, , 21, 197-221.

[90] Starks, K. J, & Burton, R. L. (1977). Preventing greenbug outbreaks. USDA Leafl. 309

[91] Starks, K. J, & Burton, R. L. (1977a). Preventing greenbug outbreaks. USDA Leafl. 309.

[92] Starks, K. J, & Burton, R. L. determining biotypes, culturing, and screening for plant resistance with notes on rearing parasitoids. USDA Tech. Bul. 1556.

[93] Stern, V. M. (1967). Control of the aphids attacking barley and analysis of yield increases in the Imperial Valley, California. J. Econ. Entomol. , 60, 485-490.

[94] Subramanyam, S, Sardesai, N, Puthoff, D. P, Meyer, J. M, Nemacheck, J. A, Gonzalo, M, & Williams, C. E. (2006). Expression of two wheat defense-response genes, Hfr-1 and Wci-1, under biotic and abiotic stresses. Plant Sci. , 170, 90-103.

[95] Ullah, F, & Peters, D. C. (1996). Sexual reproduction capabilities of greenbugs (Ho-
      moptera: Aphididae). J Kans Entomol Soc , 69, 153-159.

[96] Ulrich, W, Czarnecki, A, & Kruszynski, T. (2004). Occurrence of pest species of the
      genus *Oulema* (Coleoptera: Chrysomelidae) in cereal fields in Northern Poland. Elec-
      tronic J. Polish Agric. Uni. 7:4.

[97] Vickerman, G. P, & Wratten, S. D. (1979). The biology and pest status of cereal aphids
      (Hemiptera: Aphididae) in Europe: A review. Bull. Entomol. Res. , 69, 1-32.

[98] Voss, T. S, Kieckhefer, R. W, Fuller, B. F, Mcleod, M. J, & Beck, D. A. (1997). Yield
      losses in maturing spring wheat caused by cereal aphids (Homoptera: Aphididae)
      under laboratory conditions. J. Econ. Entomol. , 90, 1346-1350.

[99] Walker, C. B, & Peairs, F. B. (1998). Influence of grazing on Russian wheat aphid (Ho-
      moptera: Aphididae) infestations in winter wheat. In: S. S. Quisenberry and F. B.
      Peairs (eds.), A response model for an introduced pest-the Russian wheat aphid.
      Thomas Say Publ. Entomol., Entomol. Soc. Am., Lanham, MD., 297-303.

[100] Walker, C. B, Peairs, F. B, & Harn, D. (1990). The effect of planting date in southeast
      on Russian wheat aphid infestations in winter wheat. In: G. Johnson (ed.), Proc. 4th
      Russian Wheat Aphid Workshop, October 10-12. Ext. Serv., Montana State Univ.,Bo-
      zeman, MT., 54-62.

[101] Walters, M. C. (1984). Progress in Russian wheat aphid (*Diuraphis noxia* Mordw.) re-
      search in the Republic of South Africa. Tech.l Commun. 191, Dept. Agr., Republic of
      South Africa.

[102] Webster, J. A, & Amosson, S. (1995). Economic impact of the greenbug in the western
      United States: Great Plains Agr. Council Publ. 155., 1992-1993.

[103] Webster, J. A, Smith, D. H, & Lee, C. (1972). Reduction in yield of spring wheat
      caused by cereal leaf beetles. J. Econ. Entomol. , 65, 832-835.

[104] Wellso, S. G, & Hoxie, R. P. (1981). Cereal leaf beetle pupation under controlled tem-
      peratures and relative humidities. Environ. Entomol.10:58.

[105] Weng, Y, & Lazar, M. D. (2002). Amplified fragment length polymorphism- and sim-
      ple sequence repeat-based molecular tagging and mapping of greenbug resistance
      gene Gb3 in wheat. Plant Breeding , 121, 218-223.

[106] Weng, Y. Q, Azhaguvel, P, Michels, G. J, & Rudd, J. C. (2007). Cross-species transfer-
      ability of microsatellite markers from six aphid (Hemiptera: Aphididae) species and
      their use for evaluating biotypic diversity in two cereal aphids. Insect Mol. Biol. , 16,
      613-622.

[107] Wiktelius, S, & Pettersson, J. (1985). Simulations of bird cherry-oat aphid population
      dynamics: A tool for developing strategies for breeding aphid-resistant plants. Ag-
      ric.Ecosyst. Environ. , 14, 159-170.

[108] Wilktlius, S. (1987). The role of grasslands in the yearly life-cycle of *Rhopalosiphum pa-di* (Homoptera: Aphididae) in Sweden.-*Ann. appl. Biol. 10*, 9-15.

[109] Williams, C. E, Collier, N, Sardesai, C. C, Ohm, H. W, & Cambron, S. E. (2003). Phe-notypic assessment and mapped markers for H31, a new wheat gene conferring re-sistance to Hessian fly (Diptera: Cecidomyiidae). Theoretical and Applied Genetics, , 107, 1516-1523.

[110] Wilson, M. C, & Shade, R. E. (1966). Survival and development of larvae of the cereal leaf beetle, Oulema melanopa (Coleoptera: Chrysomelidae), on various species of Gramineae. Annals of the Entomological Society of America. , 59, 170-173.

[111] Wilson, M. C, & Shade, R. E. (1964). The influence of various Gramineae on weight gains of postdiapause adults of the cereal leaf beetle, *Oulema melanopa*. Ann. Entomol. Soc. Am. , 57, 659-661.

[112] Wood, E. A. Jr. (1961). Biological studies of a new greenbug biotype. J. Econ. Ento-mol. , 54, 1171-1173.

[113] Yu, G. T, Williams, C. E, Harris, M. O, Cai, X, Mergoum, M, & Xu, S. S. (2010). Devel-opment and Validation of Molecular Markers closely linked to Hessian Fly-resistance Gene H32 in Wheat. Crop Science, , 50, 1325-1332.

[114] Zaayman, D, Lapitan, N. L. V, & Botha, A-M. (2008). Dissimilar molecular defense re-sponses are elicited in *Triticum aestivum* L. after infestation by different *Diuraphis nox-ia* (Kurjumov) biotypes. Plysiol. Plantarum. , 136, 209-222.

Permissions

The contributors of this book come from diverse backgrounds, making this book a truly international effort. This book will bring forth new frontiers with its revolutionizing research information and detailed analysis of the nascent developments around the world.

We would like to thank Aakash Goyal and Muhammad Asif, for lending their expertise to make the book truly unique. They have played a crucial role in the development of this book. Without their invaluable contribution this book wouldn't have been possible. They have made vital efforts to compile up to date information on the varied aspects of this subject to make this book a valuable addition to the collection of many professionals and students.

This book was conceptualized with the vision of imparting up-to-date information and advanced data in this field. To ensure the same, a matchless editorial board was set up. Every individual on the board went through rigorous rounds of assessment to prove their worth. After which they invested a large part of their time researching and compiling the most relevant data for our readers. Conferences and sessions were held from time to time between the editorial board and the contributing authors to present the data in the most comprehensible form. The editorial team has worked tirelessly to provide valuable and valid information to help people across the globe.

Every chapter published in this book has been scrutinized by our experts. Their significance has been extensively debated. The topics covered herein carry significant findings which will fuel the growth of the discipline. They may even be implemented as practical applications or may be referred to as a beginning point for another development. Chapters in this book were first published by InTech; hereby published with permission under the Creative Commons Attribution License or equivalent.

The editorial board has been involved in producing this book since its inception. They have spent rigorous hours researching and exploring the diverse topics which have resulted in the successful publishing of this book. They have passed on their knowledge of decades through this book. To expedite this challenging task, the publisher supported the team at every step. A small team of assistant editors was also appointed to further simplify the editing procedure and attain best results for the readers.

Our editorial team has been hand-picked from every corner of the world. Their multi-ethnicity adds dynamic inputs to the discussions which result in innovative outcomes. These outcomes are then further discussed with the researchers and contributors who give their valuable feedback and opinion regarding the same. The feedback is then collaborated with the researches and they are edited in a comprehensive manner to aid the understanding of the subject.

Apart from the editorial board, the designing team has also invested a significant amount of their time in understanding the subject and creating the most relevant covers. They scrutinized every image to scout for the most suitable representation of the subject and create an appropriate cover for the book.

The publishing team has been involved in this book since its early stages. They were actively engaged in every process, be it collecting the data, connecting with the contributors or procuring relevant information. The team has been an ardent support to the editorial, designing and production team. Their endless efforts to recruit the best for this project, has resulted in the accomplishment of this book. They are a veteran in the field of academics and their pool of knowledge is as vast as their experience in printing. Their expertise and guidance has proved useful at every step. Their uncompromising quality standards have made this book an exceptional effort. Their encouragement from time to time has been an inspiration for everyone.

The publisher and the editorial board hope that this book will prove to be a valuable piece of knowledge for researchers, students, practitioners and scholars across the globe.

List of Contributors

Mohamed S. Alhammadi and Ali M. Al-Shrouf
Research & Development Division, Abu Dhabi Food Control Authority, United Arab Emirates

Kazuo Oki and Keigo Noda
The University of Tokyo, Japan

Koshi Yoshida and Issaku Azechi
Ibaraki University, Japan

Masayasu Maki and Koki Homma
Kyoto University, Japan

Chiharu Hongo
Chiba University, Japan

Hiroaki Shirakawa
Nagoya University, Japan

Santosh Kumar and Arvind H. Hirani
Department of Plant Science, University of Manitoba, Winnipeg, Manitoba, Canada

Muhammad Asif
Agricultural, Food and Nutritional Science, Agriculture/Forestry Centre, Univ. of Alberta, Edmonton, AB, Canada

Aakash Goyal
Bayer Crop Science, Saskatoon, Canada

E. T. Gwata
University of Venda, Department of Plant Production, Thohoyandou, Limpopo, South Africa

H. Shimelis
University of KwaZulu-Natal, African Center for Crop Improvement, Scottsville, South Africa

Juan Hirzel and Pablo Undurraga
Instituto de Investigaciones Agropecuarias INIA, Centro Quilamapu, Chillan, Chile

Sikander Khan Tanveer, Xiaoxia Wen and Yuncheg Liao
College of Agronomy, Northwest A&F University Yangling, Shaanxi, China

Atif Kamran
Seed Centre, Department of Botany, University of the Punjab, Lahore, Pakistan

Syed Bilal Hussain
Faculty of Agricultural Sciences & Technology, Bahauddin Zakariya University, Multan, Pakistan

Mukhtar Ahmad
Department of Agronomy, PMAS Arid Agriculture University Rawalpindi, Pakistan

Arvind Hirani
Department of Plant Science, University of Manitoba, Winnipeg, Manitoba, Canada

Printed in the USA
CPSIA information can be obtained
at www.ICGtesting.com
JSHW011357221024
72173JS00003B/317